Nazanin Elham

Flexible Decoder for LDPC Codes

Nazanin Elhami-Khorasani

Flexible Decoder for LDPC Codes

A technique to attain desirable trade-off
between performance and complexity

VDM Verlag Dr. Müller

Imprint

Bibliographic information by the German National Library: The German National Library lists this publication at the German National Bibliography; detailed bibliographic information is available on the Internet at http://dnb.d-nb.de.

Cover image: www.purestockx.com

Publisher:
VDM Verlag Dr. Müller Aktiengesellschaft & Co. KG , Dudweiler Landstr. 125 a, 66123 Saarbrücken, Germany,
Phone +49 681 9100-698, Fax +49 681 9100-988,
Email: info@vdm-verlag.de

Zugl.: Montreal,Concordia University, Diss., 2007

Produced in USA and UK by:
Lightning Source Inc., La Vergne, Tennessee, USA
Lightning Source UK Ltd., Milton Keynes, UK

ISBN: 978-3-8364-5823-8

ACKNOWLEDGEMENTS

I would like to express my sincere gratitude to my supervisor, Dr.Yousef R. Shayan for his constructive comments and valuable support throughout this work. This thesis is the result of his supervision, thoughtful guidance and encouragement.

I extend my thanks to the faculty and staff of the Department of Electrical and Computer Engineering at Concordia University, and my friends who helped and encouraged me throughout my studies.

Finally, my special thanks goes to my parents and my sister for their endless love and support. I am truly grateful to them for their patience and sacrifices throughout these years.

Dedicated to my parents

TABLE OF CONTENTS

LIST OF TABLES

LIST OF FIGURES

LIST OF ACRONYMS

APP   A Posteriori Probabilities

AWGN   Additive White Gaussian Noise

BER   Bit Error Rate

BIBD   Balanced Incomplete Block Design

BP   Belief Propagation

BPSK   Binary Phase Shift Keying

GA   Gallager A

GF   Galois Field

H_{ST} Hybrid Switch-Type

H_{TI} Hybrid Time-Invariant

LDPC   Low-Density Parity-Check

MER   Message Error rate

MS   Min-Sum

PDF   Probability Density Function

SP   Sum-Product

SS   Sipser Spielman

ST   Switch Type

TG   Tanner Graph

UL   Utilization Level

WBF   Weighted Bit Flipping

LIST OF SYMBOLS

A   Acceptable position in Gallager matrix

b   Switching threshold in Gallager algorithm

BC_{AVE} Per bit complexity based on average number of iterations

BC_{MAX} Per bit complexity based on maximum number of iterations

c   Codeword

$\hat{c}_i$ Estimated codeword

C_i Set of row locations of ones in i^{th} column

$C_{i\backslash j}$ Set of row locations of ones in i^{th} column excluding location j

c_j Check node with index j

d_c Degree of the column

d_r Degree of the row

E   Set of edges/ Number of edges

e_i Edge

Eb/No Energy per bit per noise power spectral density

G   Generator matrix

H   Parity-check matrix

H^* New H matrix

H_{sys} Systematic H

H_1, H_2 Sparse parity-check matrix

h_{ij} Entry of the parity-check matrix

I   Iteration number

I_{AVE} Average number of iterations

I_K	Identity matrix
I_M	$H_2^{-1}H_2$
I_{MAX}	Maximum number of iterations
I_u	Iteration number for node u
j	Number of ones in columns of matrix
k_{ji}	Constant
l	Length of the cycle or closed walk
m	Length of the message or number of rows
m_0	Message from channel
M	Number of columns in diagonal form at H^*
$m_{c \rightarrow v}$	Updated message from check node to variable node
$m_{v \rightarrow c}$	Updated message from variable node to check node
n	Length of the codeword or number of columns
P	$H_2^{-1}H_1$
$q_{ji}(b)$	Messages from variable node v_i to check node c_j
r	Received vector
$r_{ji}(b)$	Messages from check node c_j to variable node v_i
R	Rate of the Code
R_j	Set of column locations of ones in j^{th} row of matrix
$R_{j \backslash i}$	Set of column locations of ones in j^{th} row excluding location i
S	Syndrome vector
T	Hybrid threshold value
U	Unacceptable position in Gallager matrix

V	Set of vertices
v	Variable node
v_i	Vertix
w_r	Weight of the row in parity-check matrix
w_c	Weight of the column in parity-check matrix
λ_i	Fraction of edges connected to variable nodes
ρ_j	Fraction of edges connected to check nodes

CHAPTER 1

INTRODUCTION

This chapter provides the literature review for the work and introduces the problems that have attracted many researchers to this topic. Furthermore, the objectives for the research and the contributions of the work are given in this chapter. Finally, the organization of the thesis is outlined.

1.1 Literature Review

Low-Density Parity-Check (LDPC) codes are a family of linear error correcting codes that have very sparse parity-check matrix. Sparse parity-check matrix H has a small number of non-zero elements in each row and column. LDPC codes were introduced by Gallager in 1962 [1, 2]. They are also known as Gallager codes in honor of Robert G. Gallager. Aside from the works in [3, 4], LDPC codes were almost forgotten until the invention of Turbo codes by Berrou *et al.* [5] and the rediscovery of the LDPC codes by Mackay [6] in 1990s. Lately, LDPC codes have been the subject of many researches and analysis because of their linear decoding complexity and near shannon limit performance [6, 7, 8, 9, 10]. These characteristics make them an attractive candidate for error correcting code applications in communication systems.

Tanner in 1980s introduced the graphical representation of the LDPC codes [3]. This idea which was generalized by wiberg *et al.* [11, 12] and Kschischang *et al.* [13, 14] has been very useful for understanding the behavior of the iterative decoding algorithms. The graphical representation of the codes are known as Tanner Graph.

Two different sets of nodes or vertices in the graph are variable nodes and check nodes. In this structure, each bit is presented as a variable node and each parity-check equation is represented as a check node.

There are a number of iterative message-passing decoding algorithms, each offering a particular trade-off between error performance and decoding complexity. Hard-decision algorithms work with single bits 1 and 0, and have a great importance for their low complexity which makes them attractive for high speed communication in the cost of performance loss. Majority decoding algorithms [15] and both Gallager algorithms are in this group [2].

Soft-decision algorithms are processing real value messages and their performance is better than hard-decision ones but they are computationally complex. One of the best performing soft-decision algorithms is "Belief Propagation (BP)" or "Sum-Product (SP) " algorithm [2, 16]. This algorithm converges to "a posteriori probabilities(APP)" on a cycle-free graph. Another algorithm with almost the same performance as BP algorithm but with less complexity is "Min-Sum (MS)", also referred to as "Max-Sum" or "Max-Product" [17, 18].

Weighted Bit-Flipping (WBF) algorithm is a soft/hard decision algorithm [19]. In WBF algorithm, check nodes update the messages based on soft-decision rules and variable nodes update the messages based on hard-decision operations. Sipser-Spielman (SS) algorithm is functioning almost the same as WBF algorithms, however its variable node update rule is slightly different [7].

Generally, LDPC codes have many cycles in their graph structure which is a characteristic of a good code. The cycles create dependencies for messages that are propagating through the graph. Therefore, iterative decoding algorithms may not converge to "a posteriori probability (APP)" solution and become suboptimal. This is a major problem especially for short block length LDPC codes (less than 10000 bits), with many short cycles in their Tanner graphs. In order to solve the above issue, we need to know "How and When" the algorithm becomes suboptimal.

Recently, many techniques have been proposed with the goal of devising an efficient and low complex decoding algorithm for LDPC codes. One of these techniques is *Scheduling* which has a great influence on performance and complexity. Schedule (update rule), is the order that messages use for propagating in the graph of the code. Conventionally, the passing of the messages follows the *Flooding Schedule* in which all the variable nodes and all the check nodes pass new messages to their neighbors at each iteration [13]. A *Vertical Shuffle Schedule* is proposed in [20] to update the information as soon as being computed which results in faster convergence than the BP algorithm. The major drawback of this schedule is that short cycles of the graph are not considered. In [21], an *Horizontal Shuffle Schedule* is designed. The same as vertical schedule, the convergence speed is increased but the short cycles of the graph are not considered.

In [22, 23, 24], Mao *et al.* suggested the *Probabilistic Schedule* and exchanged the messages according to some parameters of the graph such as short cycles. *Reliability-Based Schedule* which is presented in [25, 26] controls the message passing based on the reliability of information at each node. These two recent schedules improve the performance of decoding algorithms with cost of adding computational complexity. Later, a *Deterministic Schedule* based on the distribution of the shortest cycle and closed walk in the graph is presented in [27, 28]. This schedule does not deal with probability or random generator but its performance depends on graph structure of the code.

In another technique which is called Hybrid decoding, the decoder switches among different decoding algorithms during the iterative process in order to provide a desirable complexity/performance trade-off. Hybrid Time-Invariant and Hybrid Switch-Type decoding are two types of Hybrid decoding techniques in the literature. In Hybrid Time-Invariant technique, two sets of nodes are partitioned and nodes in different partitions perform different algorithms [29, 30, 31, 32]. In Hybrid Switch-Type technique, the decoder is allowed to choose its decoding algorithm during iterative process [33, 34, 35]. Both of these techniques are trying to increase the performance and speed of convergence and decrease the complexity of the decoder. However, the desired performance/complexity trade-off depends on the selection of algorithms and their decoding thresholds.

In brief, short cycles in the Tanner graph structure of the code is one of the main problems which has a strong effect on the performance of the code. This issue has more significance in short block length LDPC codes due to the loopy nature of their graph. In addition, message updating rules and Hybrid decoding technique have a great effect on performance and complexity of LDPC codes. Based on the above studies, motivations and objectives of this work are defined in the next section.

1.2 Motivations and Objectives

Proposed decoding algorithms in literature have two major drawbacks. First, the algorithms are not optimal in presence of short cycles in the graph structure of the code. Moreover, there is still a need for new decoding algorithms with low complexity and high performance. These made us motivated to merge Deterministic schedule and

Hybrid Switch-Type technique with the purpose of achieving both goals concurrently.

Based on the above motivation, the main objectives of this research are summarized as follows:

- Studying the effect of short cycles in the graph of the code to find out "How and When" the algorithm becomes sub-optimal,

- Improving the performance of the algorithms individually by preserving their optimality.

- Decreasing the complexity of the algorithms by reducing the average number of soft iterations and total number of iterations required for convergence.

- Designing a flexible multistage decoder for any desired LDPC code and any combination of decoding algorithms, in order to provide a desirable performance/complexity trade-off based on the communication systems need.

1.3 Contributions

In this thesis, we narrow down the scope of our work to *message-passing schedules* and *Hybrid decoding* technique. These two different techniques are merged in our methodology to fulfill the requirement of a decoder with low complexity and high performance. At the first step of the work, Deterministic node-based schedule which preserves the optimality of the algorithms is applied on Sum-Product (SP) [16] and Gallager A (GA) [2]. In order to apply the schedule, Tanner graph structure of LDPC codes is studied and a search algorithm for finding the shortest closed path of each node in the graph is developed. At the second step, the improved algorithms are used to design a two stage decoder with Hybrid switch-type technique. The proposed algorithm provides a desirable trade-off between performance and complexity based on predefined thresholds. The contributions of this work is outlined as follows:

- Tanner Graph structure for LDPC codes are studied.

  - Studying the graph structure of LDPC codes to understand the operation of Deterministic schedule and the sub-optimality cases for decoding algorithms.

  - Proposing a graph-based search algorithm to find the shortest closed walk and shortest cycle for each node of the graph.

  - Presenting and proving a lemma and a theorem used to design the search algorithm.

- Deterministic schedule and Hybrid decoding technique to provide complexity/performance trade-off are combined.

  - Applying Deterministic node-based schedule on variable nodes, check nodes or variable/check nodes of the short block length LDPC codes to preserve the optimality of the algorithms.

  - Applying Hybrid Switch-Type technique with different decoding thresholds on combination of a hard-decision *Gallager A* and a soft-decision *Sum-Product* decoding algorithms to provide a desirable performance/complexity trade-off.

- The performance of the decoding algorithms for regular and irregular LDPC codes are studied.

  - Providing simulation results for regular random structured (1200,600) LDPC code with rate 1/2, degree of the columns 3 and degree of the rows 6.

- Providing simulation results for irregular (1008, 504) LDPC code with rate 1/2, degree of the columns $\{2, 3, 4, 5, 7, 14, 15\}$ and degree of the rows $\{7, 8, 9\}$, optimized for AWGN channel.

- Complexity of the decoding algorithms for regular and irregular codes are analyzed.

- Providing statistics on the complexity of conventional and new decoding algorithms based on required average number of iterations for convergence in their waterfall region.

- Studying the complexity of the algorithms at bit level based on average number of iterations and maximum number of iterations.

- Providing average number of hard iterations and average number of soft iterations required for convergence of each algorithm.

1.4 Thesis Outline

This thesis is organized in five chapters. The introduction chapter provides a brief review of the previous research done on this topic. It also gives us a general overview of the work.

Chapter 2, provides the necessary background knowledge on LDPC codes which are used in subsequent chapters. Consequently, characteristics of LDPC codes and their presentation methods, encoding rules, iterative decoding algorithms and decoding rules which are used in subsequent chapters are introduced.

Chapter 3, defines the main theme of this works. First, sub-optimality cases

7

for decoding algorithms are investigated and a search algorithm for finding the sub-optimality of the nodes is developed. Then, the scheduling technique and hybrid decoding technique for decoding LDPC codes is introduced. Finally, the main idea of this work, which is the combination of these two techniques for trading-off between complexity and performance of current iterative decoding algorithms is described in detail.

Chapter 4, outlines the simulation system model and system requirements of this work. Simulation results for studying the performance of the new decoding technique, which is discussed in chapter 3, are provided. The simulation results are based on different LDPC codes and iterative decoding algorithms. Further more, decoding complexity and statistics of iteration numbers for different algorithms are analyzed and discussed.

Chapter 5 concludes the thesis. It highlights the contributions and achievements of this thesis. Some suggestions for the future work are also discussed in this chapter.

CHAPTER 2

LOW-DENSITY PARITY-CHECK CODES

In this chapter, the necessary background on LDPC codes and their characteristics are provided. Furthermore, the representation methods of the code, construction methods, encoding methods and the iterative decoding algorithms for LDPC codes are explained.

The organization of the chapter is as follows. In section 2.1, block and convolutional, binary and non-binary and regular and irregular LDPC codes are explained. In section 2.2, parity-check matrix and Tanner graph representation of LDPC codes are depicted. Random and structure construction for LDPC codes are explained in section 2.3. Section 2.4 describes the systematic method for encoding LDPC codes. Some general knowledge about iterative decoding algorithms are provided in section 2.5 and decoding rules for Sum-Product algorithm and Gallager A algorithm are given. The summary of the chapter is given in section 2.6.

2.1   Characteristics of LDPC codes

Low-Density Parity-Check (LDPC) codes are distinguished from conventional linear codes because of their sparse parity-check matrix which has small number of non-zero elements compared to its total number of entries. Another attractive feature of these codes is their efficient iterative decoding. The complexity of the iterative decoding algorithm is related to the number of ones in their parity-check matrix which creates the connection between check nodes and variable nodes. This complexity is low for

LDPC codes due to their sparse matrix. Moreover, the operation of iterative decoding algorithms can be described on Tanner Graph of the code. Therefore, the graphical representation of LDPC codes has a great importance. In this chapter some important characteristics of LDPC codes are explained.

2.1.1 Block and Convolutional LDPC Codes

Many researchers are working on LDPC *block* codes because of their great performance that can approach close to shannon limit and have relatively simple iterative decoder. Low-density parity-check *convolutional* codes are another category of LDPC codes, which have been proposed by Felstrom and Zigangirov in the late 1990s [36]. LDPC convolutional codes have also very good performance and use decoding algorithms similar to LDPC block codes. Their advantage over the block codes is their ability to work with arbitrary length of information bits without fragmenting them into blocks. This feature is used in packet based communication systems dealing with variable length packets like IEEE 802.3 Ethernet standards. In this thesis LDPC block codes are considered for the research.

2.1.2 Binary and Non-Binary LDPC Codes

Most of the research publications have considered LDPC codes over $GF(2)$, however they can be extended to $GF(q)$ by considering a set of non-zero weights $\in GF(q)$, forming the parity-check matrix [37]. In this thesis we use binary LDPC codes, hence the elements of parity check matrices are 0 or 1.

An (n, j, k) LDPC code is a block code of length n with j number of ones in each column and k number of ones in each row of the parity-check matrix H. The parity-check matrix consists of zeros and ones. Given information bits of length k

and codeword of length n, rate of the code R is defined by k/n which means $(n - k)$ redundant bits have been added to the message to correct the errors.

2.1.3 Regular and Irregular LDPC Codes

There are two structures for LDPC codes, *regular* and *irregular*. In regular LDPC codes, all columns of the parity-check matrix have the same number of ones and all rows have the same number of ones as well. This is shown in Fig. 2.1. The number of ones in each column can be represented by the degree of the column d_c or weight of the column w_c, and the number of ones in each row by degree of the row d_r or weight of the row w_r and $d_r = d_c(n/m)$ which gives the total number of ones in the matrix. n is the number of columns and m is the number of rows in the parity-check matrix.

$$H = \begin{bmatrix} 1 & 1 & 1 & 1 & 0 & 0 & 0 & 0 & 0 & 0 \\ 1 & 0 & 0 & 0 & 1 & 1 & 1 & 0 & 0 & 0 \\ 0 & 1 & 0 & 0 & 1 & 0 & 0 & 1 & 1 & 0 \\ 0 & 0 & 1 & 0 & 0 & 1 & 0 & 1 & 0 & 1 \\ 0 & 0 & 0 & 1 & 0 & 0 & 1 & 0 & 1 & 1 \end{bmatrix}$$

Figure 2.1: (10, 2, 4) Regular Parity-Check Matrix

If the number of ones in each row or column are not fixed, LDPC codes are called irregular. For irregular codes the variable node and check node degree distributions are defined by sequences $(\lambda_1, \lambda_2, ..., \lambda_{d_v})$ and $(\rho_1, \rho_2, ..., \rho_{d_c})$. λ_i is the fraction of edges connected to variable nodes with degree i and ρ_j is the fraction of edges connected to check nodes with degree j. d_v is the maximum degree of variable nodes and d_c is the maximum degree of check nodes. We can also show this sequence based on generating polynomials $\lambda(x) = \sum_i \lambda_i x^{i-1}$ and $\rho(x) = \sum_i \rho_i x^{i-1}$ [10, 38].

Fig. 2.1, shows the parity-check matrix of a (10, 2, 4) regular LDPC code with rate 1/2, block length 10, weight of the columns $w_c = d_v = 2$, weight of the rows $w_r = d_c = 4$ and number of information bits $k = 5$. After using generating polynomials for this code, we have $\lambda(x) = x$ and $\rho(x) = x^3$.

2.2 Representation of LDPC Codes

The graph representation of LDPC codes is analogous to their matrix representation. The parity-check matrix of LDPC code can be obtained by having the Tanner graph structure of the code. Assuming binary $m \times n$ parity-check matrix H, its entry h_{ij} is 1, if and only if, ith check node is connected to jth variable node in the graph. Conversely, a bipartite graph between n messages and m check nodes of the same binary $m \times n$ matrix can be obtained by assigning a connection between any check node and message node of the graph that has 1 entry in the parity-check matrix. A bipartite graph is a special graph where the set of vertices can be divided into two disjoint sets v and c such that every edge has one end-point in v and one end-point in c and no edge exists between vertices at the same set.

Two fundamental ways for representing LDPC codes, parity-check matrix and Tanner graph, are going to be studied in this section.

2.2.1 Parity-Check Matrix Representation

LDPC codes follow the same definition which already exists for other linear block codes [39]. A message word $m = m_1 m_2 ... m_k$ with length k can be mapped to codeword $c = c_1 ... c_n$ with length n.

$$c_1 = m_1 g_{1,1} + \cdots + m_k g_{k,1}$$

$$\vdots = \vdots \qquad\qquad (2.1)$$

$$c_n = m_1 g_{1,n} + \cdots + m_k g_{k,n}$$

We convert these equations into matrix form

$$c = m\mathbf{G} \qquad\qquad (2.2)$$

where G

$$G = \begin{bmatrix} g_{1,1} & \cdots & g_{1,n} \\ \vdots & \ddots & \vdots \\ g_{k,1} & \cdots & g_{k,n} \end{bmatrix} \qquad\qquad (2.3)$$

is a $k \times n$ full rank *generator* matrix. From the G matrix we can find $(n-k) \times n$ parity-check matrix H such that $GH^T = 0$ and any row of H is orthogonal to all rows of G. Therefore, based on Eq. 2.2 we will have $cH^T = 0.H$ is the parity-check matrix of the code which can be designed based on the degree distribution of the nodes (weight of the columns and rows).

2.2.2 Tanner Graph Representation

Tanner graph (TG) of the code gives us a better idea about the structure of decoding algorithm and gives a complete representation of the code. Tanner graph of the LDPC codes is a bipartite graph, hence there is no connection or edge between any two nodes of the same set. In Fig. 2.2, variable nodes are shown by circles and check nodes by squares which is a normal practice in the literature. The connection is based on the 0s and 1s in the parity check matrix. Check node $c_j, j = 1, 2, ...n - k$ is connected to bit node $v_i, i = 1, 2, ...n$, if the value of element h_{ji} in matrix H is 1.

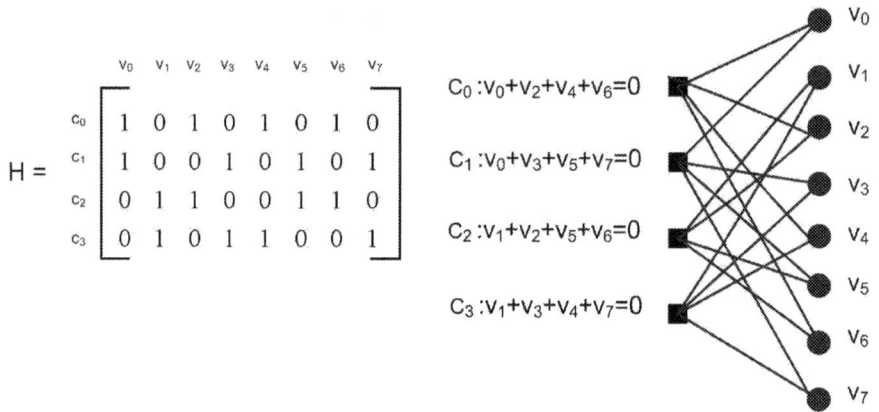

Figure 2.2: Representation of LDPC Codes

We can see in Fig. 2.2 that there are m or $(n - k) = 4$ check nodes on the left side and $n = 8$ variable nodes on the right side of the graph. On the other hand the $m = 4$ rows of H represents the m check nodes and the $n = 8$ columns of H represents the n variable nodes. As an example, c_0 is connected to v_0, v_2, v_4, v_6 by the solid lines in the graph and in the first row of the matrix $h_{00}, h_{02}, h_{04}, h_{06}$ are equal to one and $h_{01} = h_{03} = h_{05} = h_{07} = 0$. The presented Tanner graph is regular and each variable node has 2 edge connections or in other words holds degree of 2 and each check node has 4 edge connections and holds degree of 4. In the parity-check matrix weight of the columns is 2 and weight of the rows is 4, respectively.

2.3   Construction of LDPC codes

Construction of LDPC codes is equivalent to design of parity-check matrix H. The design of H requires the knowledge of all the parameters, like the desired block length of the code, degree distributions, rate of the code and other characteristics.

In the design of parity-check matrix we should try to increase the girth (girth is the length of the shortest cycle in graph), which will be discussed more in the next chapter. On the other hand, for high minimum distance, we should try to increase the sparseness of the matrix.

By considering these factors, there are many different ways for constructing parity-check matrices and designing LDPC codes. These methods are generally defined as random and structure construction.

2.3.1   Random Construction

Random construction was the first method used to design LDPC codes. Random construction does not have many constraints and can be easily applied for different combination of parameters, however it does not guarantee the size of the girth. For a desired girth size, random constructions may be used with some added constraints. Mackay and Gallager random construction methods are explained in this section.

a) Mackay Construction :

Mackay designed various random construction methods [8] and confirmed near Shannon limit performance of LDPC codes [6]. The major drawback of his codes is the lack of structure which results in high complexity encoding. Most of his designs are creating regular LDPC codes based on bipartite graphs. Different Mackay construction methods are explained as follows.

- Construction 1A

  An m-row by n-column matrix, created randomly with fixed number of ones in each column and weight per row as uniform as possible. In this matrix the overlap between any two columns is not more than 1. Therefore the matrix does not have cycles with length 4. Since the graph of the code is bipartite, the length of the cycles must be even. This results in girth size equal to 6.

- Construction 2A

  In an m-row by n-column matrix, $m/2$ of the columns are designed with weight 2 and there is no overlap between any two columns. The rest of the columns are constructed randomly with weight 3. Weights of the rows are as uniform as possible and the overlap between any two columns of the entire matrix is not greater than 1.

- Construction 1B and 2B

  An m by n matrix is constructed using methods 1A or 2A. Then small number of columns of this matrix are chosen and deleted randomly. Hence, the bipartite graph corresponding to the matrix has no short cycles of less than some length. By reducing the possibility of having short cycles in the matrix the performance of the code is improved.

b) Gallager Construction:

Gallager in his thesis designed an ensemble of LDPC codes, then applied some constraints to design the matrix without having cycles of length 4 [2].

To construct the matrix, first we define an (n, j, k) parity-check matrix, a matrix with n columns that has j ones in each column and k ones in each row and zero

16

in other locations. For constructing an ensemble of (n, j, k) matrices, we use an example taken from Gallager's dissertation as $n = 20$, $j = 3$ and $k = 4$. The matrix is divided into j submatrices and each submatrix contains a single 1 in each column. The first division of these submatrices puts the 1's in descending order. It means the ith row contains ones in columns $(i - 1)k + 1$ to ik as it is shown in Fig. 2.3. The other submatrices are column permutation of the first division. Gallager defines the ensemble of (n, j, k) codes as the ensemble resulting from random permutations of the columns of the bottom $(j - 1)$ submatrices of a matrix with equal probability assigned to each permutation.

```
1 1 1 1 0 0 0 0 0 0 0 0 0 0 0 0 0 0 0 0
0 0 0 0 1 1 1 1 0 0 0 0 0 0 0 0 0 0 0 0
0 0 0 0 0 0 0 0 1 1 1 1 0 0 0 0 0 0 0 0
0 0 0 0 0 0 0 0 0 0 0 0 1 1 1 1 0 0 0 0
0 0 0 0 0 0 0 0 0 0 0 0 0 0 0 0 1 1 1 1
1 0 0 0 1 0 0 0 1 0 0 0 1 0 0 0 1 0 0 0
0 1 0 0 0 1 0 0 0 1 0 0 0 0 0 0 1 0 0 0
0 0 1 0 0 0 1 0 0 0 0 0 0 1 0 0 0 1 0 0
0 0 0 1 0 0 0 0 0 0 1 0 0 0 1 0 0 0 1 0
0 0 0 0 0 0 0 1 0 0 0 1 0 0 0 1 0 0 0 1
1 0 0 0 0 1 0 0 0 0 0 0 1 0 0 0 0 1 0 0
0 1 0 0 0 0 1 0 0 0 1 0 0 0 0 1 0 0 0 0
0 0 1 0 0 0 0 1 0 0 0 0 1 0 0 0 0 0 1 0
0 0 0 1 0 0 0 0 1 0 0 0 0 1 0 0 1 0 0 0
0 0 0 0 1 0 0 0 0 1 0 0 0 0 1 0 0 0 0 1
```

Figure 2.3: Example of Gallager Matrix for n=20, j=3 and k=4

Gallager also constructs parity-check matrices without cycles of length 4. We refer to the previous example ($n = 20, j = 3, k = 4$) and describe the procedure as follows. First we consider an nj/k by n matrix. The matrix should be divided into $jk = 12$ submatrices, each with $n/k = 5$ rows and columns. The first row and column of submatrices are identity matrices. The rest of the submatrices contain the letter U in each main diagonal and A in other positions. We have to find 5 non-zero elements

17

in each submatrix in a way that the overlap between any two columns of H matrix does not exceed one. The letter U represents unacceptable position, since it can create an overlap of more than one between any two columns of the matrix.

Then, we choose a submatrix with U and A elements. In the first row of the submatrix, we choose a position of A and change it into 1. Hence, the rest of the positions in the same row and the same column of the chosen position at the targeted submatrix can not be 1. Also, the same position in some other submatrices can not be 1 and should be changed to zero. We should continue this procedure until we fill in all the positions and do the transform in other rows of the submatrix and extend to other submatrices as well. The matrix made by the above technique does not have more than one overlap between any two columns, therefore the parity-check matrix has no cycle of length 4 or less. In the next chapter, we will use a simple search method to find and remove all the 4 cycles in parity-check matrix.

There are many other random constructions used in literature like Bit filling construction [40] and Average girth distribution based construction [41]. The former structure is a search method for finding LDPC codes with large girth and the latter searches for good short length LDPC codes based on the average of the girth distribution of the code.

2.3.2 Structured Construction

Since the random structure requires a lot of memory space to keep the non-zero elements in the random parity check matrix, construction of structured LDPC codes is applied on many applications to reduce the hardware cost and simplify the encoding/decoding system. Cyclic and quasi-cyclic LDPC codes are two examples of structured designs. There are many ways to construct them but three typical methods are quasi cyclic LDPC codes based on finite geometries [42, 19], balanced incomplete block design (BIBD) and disjoint different sets [43].

2.4   Encoding of LDPC codes

After designing the H matrix, the generator matrix G can be found easily from H. One of the weak points of LDPC codes is their encoding complexity. As a result many different encoding schemes have been suggested for LDPC codes to reduce the complexity [44, 45]. In this work, systematic encoding is used and described in details.

2.4.1   Systematic Encoding

In systematic encoding Gaussian elimination is used to find the generator matrix G from corresponding H matrix [8]. The complexity of the calculation is high in this technique, especially when the length of the codeword is increased. There are several other encoding algorithms with lower complexities but we use the systematic encoding in this thesis. For short block lengths, systematic encoding is preferable because the size of the matrix is not big and method results low complex decoding. Systematic Generator matrix G can be derived from H through following steps:

1. Choose a regular or irregular H matrix.

2. Reorder the columns of H in a way that the first M columns of new H matrix (H^*) have all one in the diagonal of matrix.

3. Apply Gaussian Elimination to H^* in order to get systematic H matrix.

4. Generator matrix G is the transpose of systematic H matrix and can be derived easily.

Now, we assume that the parity check matrix has a formation like

$$H = \left[\; H_1 \mid H_2 \;\right].\tag{2.4}$$

H_1 and H_2 are two very sparse matrices and the rows in H are linearly independent. Matrix H_1 is a rectangular $k \times (n-k)$ and matrix H_2 is a square $k \times k$ invertible matrix. We can reorder the columns of H matrix, in order to get invertible H_2. The equivalent parity-check matrix is

$$H^* = H_2^{-1}H = H_2^{-1}\left[\; H_1 \mid H_2 \;\right] = \left[\; P|I_M \;\right]\tag{2.5}$$

where

$$P = H_2^{-1}H_1 \ and \ I_M = H_2^{-1}H_2.\tag{2.6}$$

The generator matrix of the LDPC code will have the following format

$$G^T = \left[\begin{array}{c} I_k \\ P \end{array}\right] = \left[\begin{array}{c} I_k \\ H_2^{-1}H_1 \end{array}\right]\tag{2.7}$$

and

$$c = G^T k.\tag{2.8}$$

20

I_k is an $k \times k$ identity matrix, c is a codeword and k is information message. Since,

$$H \times G^T = H^* \times G^T = 0 \qquad (2.9)$$

it can be implied from above equations that, we are able to apply systematic Generator matrix G to the encoder and sparse parity-check matrix H to the decoder. The reason for using the sparse parity-check matrix is reducing the complexity of the decoder.

2.5 Decoding of LDPC Codes

Let c be the transmitted codeword through a noisy communication channel and r be the received codeword at the input of the demodulator. The codeword c is an encoded message using $c = G^T u$. Generator matrix G is in systematic form. Source vector u has length k (information bits) and is encoded into a transmitted vector c. The received vector $r = (G^T u \oplus n)$, is the combination of noise n and the transmitted vector. Since the decoder does not know the noise and transmitted message patterns, it faces the task of finding the most likely message vector u that is sent through the channel.

Decoder uses the syndrome decoding technique to compute the following equation. Syndrome of received vector r will be calculated as

$$S = rH^T. \qquad (2.10)$$

If r is a codeword, the syndrome will be zero and we accept r as a transmitted codeword. If r is not a codeword, the syndrome is not zero, hence we are facing error in the received vector. The usefulness of the above equation is its dependence only on the noise vector and the parity-check matrix and it does not have any relation with the transmitted codeword. It means,

$$S = rH^T \Rightarrow S = (c \bigoplus n)H^T \Rightarrow S = cH^T \bigoplus nH^T \qquad (2.11)$$

where $cH^T = 0$ and $S = nH^T$. This is the relation between the noise and syndrome vector. Now we can get estimate of the transmitted codeword from $c = r - n$. But finding the noise vector is not always an easy task, hence many decoding algorithms are proposed to solve this problem.

2.5.1 Iterative Decoding Algorithms

Message updating behavior of iterative decoding algorithms is depicted in Fig. 2.4. In each iteration of the iterative decoding, the information from the channel (intrinsic information) and the information from previous iterations coming from different nodes (extrinsic information) are used to obtain a better knowledge about the transmitted message. LDPC codes can be decoded by different iterative decoding algorithms.

Generally all the iterative decoding algorithms follow 4 steps. At the initialization step in iteration $I = 0$, an initial or local message will be assigned to each variable node. The value of this message will be based on the observations of the output of the channel.

From the first iteration $(I \geq 1)$ in the second step, the initial message will be passed from variable nodes to check nodes through the edges of the Tanner Graph

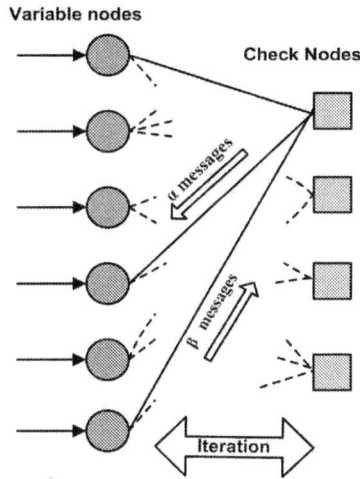

Figure 2.4: Message Updating in Iterative Decoding Algorithms

as it is shown in Fig. 2.4 by β messages. At the third step, check nodes process the incoming messages from variable nodes based on the decoding schedules and send new estimates to the variable nodes. Variable nodes process the incoming messages from check nodes (α messages) which are updated in previous iterations and send back the updated value to the check nodes. Messages sent along the edges are independent from each other.

At step 4 in each iteration, at variable nodes, algorithm approximates the codeword from the probabilistic information and makes a hard decision. The algorithm stops if $cH^T = 0$ or maximum number of iterations is reached. Otherwise, algorithm continues the iterations.

The main theme of this work, which is developed in chapter 3, is applied on two different algorithms. The first algorithm is a soft-decision algorithm called "$Sum - Product(SP)$", which is the most important probabilistic decoding algorithm used for LDPC codes. It minimizes the probability of decoding error for a given code and is optimal when the TG of the code is cycle-free but the draw-back of this algorithm is its computational complexity . The other algorithm is "$Gallager A(GA)$" algorithm. This algorithm is especially important for its very simple implementation and low complexity but its draw-back is the low performance.

2.5.2 Sum-Product Algorithm

Before explaining steps of the decoding algorithm, notations used in the algorithm are given as follows:

- $r_{ji}(b)$ is the message that will be sent from check node c_j to variable node v_i and $b \in \{0,1\}$. $r_{ji}(0)$ and $r_{ji}(1)$ are representing the probabilities of receiving 0 or 1.

- $q_{ji}(b)$ is the message that will be sent from variable node v_i to check node c_j. $q_{ji}(0)$ and $q_{ji}(1)$ are representing the probabilities of receiving 0 or 1.

- R_j is the set of column locations of ones in the jth row of matrix.

- $R_{j\setminus i}$ is the set of column locations of ones in the jth row of matrix, excluding location i.

- C_i is the set of row locations of ones in the ith column.

- $C_{i\setminus j}$ is the set of row locations of ones in the ith column of matrix, excluding location j.

Following steps represent Sum-Product algorithm in probabilistic domain [2, 13]:

Step 1: At the initialization stage, each variable node sends a message along each

24

of its outgoing edges to indicate the probability of having "1" at that node when the output of the channel for that bit is given.

$$q_{ji}(0) \quad = \quad 1 - p_i = P_r(x_i = +1|y) = \frac{1}{1 + e^{-2y_i/\sigma^2}} \tag{2.12}$$

$$q_{ji}(1) \quad = \quad p_i = P_r(x_i = -1|y) = \frac{1}{1 + e^{2y_i/\sigma^2}} \tag{2.13}$$

Step 2: Check nodes calculate their response messages using check node update rule.

$$r_{ji}(0) \quad = \quad 1/2 + 1/2 \prod_{i \in R_{j \setminus i}} (1 - 2q_{ji})(1) \tag{2.14}$$

$$r_{ji}(1) \quad = \quad 1 - r_{ji}(0) \tag{2.15}$$

Step 3: Variable nodes update their response messages to check nodes using variable node update rule .

$$q_{ji}(0) \quad = \quad K_{ji}(1 - p_i) \prod_{\acute{j} \in C_i} r_{\acute{j}i}(0) \tag{2.16}$$

$$q_{ji}(1) \quad = \quad K_{ji}p_i \prod_{\acute{j} \in C_{i \setminus j}} r_{\acute{j}i}(1) \tag{2.17}$$

K_{ji} are chosen in a way to ensure that $q_{ji}(0) + q_{ji}(1) = 1$.

Step 4: The following equations must be calculated for all the i's. Variable nodes update their current estimate by calculating the probabilities for "0" and "1".

$$Q_i(0) \quad = \quad K_i(1 - p_i) \prod_{j \in C_i} r_{ji}(0) \tag{2.18}$$

$$Q_i(1) \quad = \quad K_i p_i \prod_{j \in C_i} r_{ji}(1). \tag{2.19}$$

25

K_i is chosen in a way to ensure that $Q_i(0) + Q_i(1) = 1$.

Step 5: After calculating the probabilities at the previous step for every row index i, variable node compares the probabilities and votes for the larger value.

$$\hat{c}_i = \begin{cases} 1 & \text{If} \quad Q_i(1) > 0.5 \\ 0 & \text{else} \end{cases} \qquad (2.20)$$

At this point, if the estimated codeword satisfies the parity-check equation the algorithm stops. Otherwise, we should go back to step 2 until we reach the maximum number of iterations.

2.5.3 Gallager A Algorithm

Gallager A is a hard-decision algorithm. Therefore, messages passing through the edges are $\{0, 1\}$ and no soft information is used. Decoding steps for Gallager A algorithm are given as follows:

Step 1: At the initialization stage, all the variable nodes will be initialized with the messages they receive from the channel.

$$m_{v \to c} = m_j^{(0)} \qquad (2.21)$$

Step 2: Check nodes update the variable node v using the modulo-two sum of all the messages coming from other variable nodes except variable node v.

$$m_{c \to v} = \bigoplus_{y \in n(c) - v} m_{y \to v} \qquad (2.22)$$

26

$\oplus$ in this equation represents modulo-two sum of binary messages.

Step 3: The outgoing message at variable node v is the same as the intrinsic message which is coming from the channel unless all the extrinsic messages from neighboring check nodes disagree with the intrinsic message. Therefore, the outgoing message will be flipped to be the same as extrinsic message.

$$m_{v \to c} = \begin{cases} m_0 & If & \exists y \in n(v) - c : m_{y \to v} = m_0 \\ \overline{m_0} & Otherwise \end{cases} \qquad (2.23)$$

m_0 is the intrinsic message and $\overline{m_0}$ is the complement of this binary message.

Step 4: Hard decision at variable nodes is based on the majority of votes at previous step. The algorithm stops if the estimated decision is a codeword or maximum number of iterations is reached.

2.6 Summary

In this chapter, a general background for LDPC codes and iterative decoding algorithms were provided. In section 2.1, block and convolutional, binary and non-binary and regular and irregular LDPC codes were explained. In this thesis, binary low-density parity-check block codes are used. The results of the experiments are based on both regular and irregular codes.

Matrix and Graph representation methods for LDPC codes were depicted in the second section. Tanner Graph structure of the codes will be studied further in the next chapter. In the third section, random and structured construction of LDPC codes were explained. The regular code used in this work is constructed randomly and the irregular code is based on structured method.

Systematic encoding of LDPC codes and iterative decoding algorithms were explained in section 2.4. Although the complexity of calculations in systematic encoding is high for large block lengths, they result in low complex decoding. In this thesis, systematic encoding is applied on short block length LDPC codes. Some general knowledge about iterative decoding algorithms and decoding rules for Sum-Product and Gallager A algorithms were provided in section 2.5.

CHAPTER 3

MULTISTAGE SCHEDULED DECODER FOR
LOW-DENSITY PARITY-CHECK CODES

In this chapter, the main theme of this work is defined. The main idea, which is the combination of Deterministic schedule and Hybrid Switch-Type technique for trading-off between complexity and performance of current iterative decoding algorithms is introduced. Also, a search algorithm to find the length of the shortest closed path for each node in the graph structure of the codes is developed.

The organization of the chapter is as follows. In section 3.1, structure of Tanner Graph is studied and some definitions on graphs are given. In section 3.2, sub-optimality cases for decoding algorithms and drawback of conventional algorithms are indicated. In section 3.3, Deterministic node-based schedule is explained. This schedule preserves the optimality of the decoding algorithm. Section 3.4, gives a comprehensive idea about the Hybrid decoding and its different types. The search strategy which is designed to find the girth and shortest closed walk of the nodes in graph is proposed in section 3.5. Finally, in section 3.6 the main idea and the advantages of the new decoder are given. Summary of the chapter is given in section 3.7.

3.1 Structure of the Tanner Graph

Knowledge about the graphs is required for the rest of this chapter. Therefore, before representing the graph-based algorithm, Tanner graph structure of the codes are studied [28, 46, 47]. Based on the definitions given below, the sub-optimality cases of the decoding algorithms are explained in the next section. In this work, the graphs are un-directional and the direction of the cycles and walks are not important as will be described later.

Definition 1: A graph G consists of a set of *vertices* $V = \{v_1, v_2, \cdots\}$ and a set of *edges* $E = \{e_1, e_2, \cdots\}$ such that each edge e_k is identified with an unordered pair of vertices (v_i, v_j). In other words, graph is a finite set of vertices (*nodes*) connected by links which are called edges. The graph can be denoted by $G(V, E)$. Fig. 3.1 is an example of a graph with 5 vertices and 6 edges.

Definition 2: The *degree* of a vertex is the number of edges which have that vertex as an end point. For example, in Fig. 3.1, degree of the nodes v_2 and v_5 is 3.

Definition 3: *Walk* is a sequence of edges, one following the other. If we have two vertices v_1 and v_2 in graph G, we define the walk (v_1, v_2) as a sequence of vertices and edges beginning from v_1 and ending with v_2 such that vertices and edges are incident. The number of edges in the walk is the length of the walk. For example in Fig. 3.1, (v_1, v_2, v_3) is a walk from v_1 to v_3 with length 2 and the walk with the form $(v_2, v_3, v_4, v_5, v_1, v_2)$ is a *closed walk*. A walk is called closed walk, if the start node and the end node are the same. *Path* is also a rout consisting of connected edges from start node to the end node.

Definition 4: A walk is *nontrivial*, if it has at least one edge that is traversed only once.

Definition 5: A *cycle* is a closed walk in which all the nodes except for the start and the end nodes are distinct. In Fig. 3.1, $(v_2, v_3, v_4, v_5, v_2)$ is a cycle and a nontrivial closed walk. In other words, cycle is a special nontrivial closed walk.

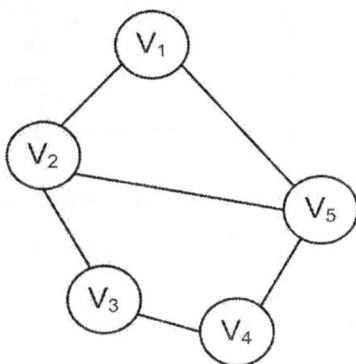

Figure 3.1: Structure of a Graph with 5 Nodes and 6 Edges

Definition 6: If any two vertices in a graph are connected by an edge, the graph is called *connected* graph.

Definition 7: A graph $G(V, E)$ is called *bipartite* if its vertex set can be partitioned into two subsets v_1 and v_2 such that no two vertices in the same subset are connected. For example, if (v_1, v_2) is an edge, its end points belong to different subsets. Fig. 3.2 represents a bipartite graph.

Definition 8: *Girth* of a graph is the length of the shortest cycle in the graph. In a bipartite graph, the shortest possible cycle has length 4. Hence, the girth of a bipartite graph is 4. Girth of a node is the shortest cycle passing through that node.

Definition 9: *Tree* is a special type of bipartite graph. In other words, a connected graph in which there is only one path connecting each pair of vertices is called tree. Tree does not have any cycle in its structure. Given two vertices v_1 and

31

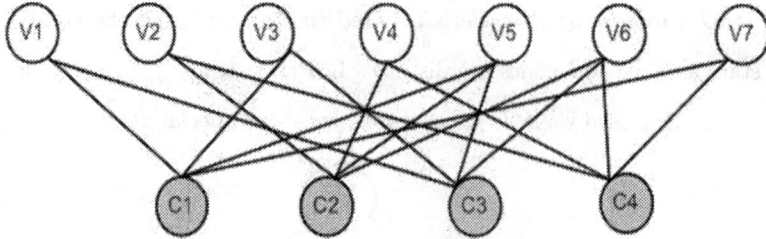

Figure 3.2: Structure of a Bipartite Graph

v_2 of a tree, there is a unique (v_1, v_2) path in the tree and deletion of any edge makes the tree disconnected. In Fig 3.3, part of a tree structure is depicted. The start node at the top of the tree is usually named as the root of the tree or *parent* node. Each node in the tree which lies along the path from a child node to the root of the tree can become parent node. For example, in Fig. 3.3 node b is the parent node for the children d and e.

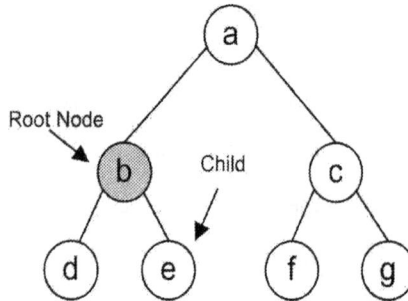

Figure 3.3: Structure of a Tree

Definition 10: Given a graph G with $V(G) = \{v_1, v_2, ..., v_n\}$, we define the $n \times n$ *adjacency matrix* H as follows. Element of the matrix h_{ij} is entry 1, if (v_i, v_j) is an edge in the graph and h_{ij} is 0, if v_i and v_j are not connected.

In this work, cw_v and g_v denote the length of the shortest nontrivial closed walk and the girth passing through node v, respectively. Vertices of the graphs are commonly called *nodes* and edges do not have direction. Based on the above definitions, the sub-optimality cases are studied in the next section.

3.2   Sub-Optimality of Decoding Algorithms

Sum-Product algorithm passes the probability of messages through the graph of the code to find the most likely message. For a cycle free Tanner graph, all the incoming messages to a node are independent of each other. Hence, Sum-Product algorithm results in "a posteriori probability (APP)". On the other hand, it is proved that the codes without cycle do not have good performance. As a result, LDPC codes contain numbers of loops in their graph structure.

In a graph with cycles and closed walks, messages that are passing through cycles create dependency and the incorrect value of a node will propagate through the graph and will be returned back to itself. At this case, the algorithm needs to be repeated (*iteration*), with the goal of correcting the message and resulting in convergence. However, the first time that dependency occurs, algorithm becomes sub-optimal and APP result is not guaranteed any more. Other decoding algorithms may lose their optimality in the same way. When the outgoing message of the node is not optimal, the *node* loses its optimality as well. The unreliable message returns to the node via a cycle or a closed walk and it can effect on the optimality of either variable node or check node. There are two kinds of dependencies for messages that are propagating in the graph [28].

Case 1: Dependency between incoming messages to a variable node and its initial (*local*) weight.

Example: In Fig. 3.4 (a), the initial message of *variable* node v_1 will propagate in the graph and return to v_1 through the cycle after 3 iterations. In the third iteration, the incoming messages to node v_1 which are coming from edges (c_2, v_1) and (c_3, v_1) are dependent on the local message of v_1. As a result, the outgoing message of v_1 in the next iteration to check node c_2 is not optimal any more due to the dependency between incoming message from check node c_3 and local message of variable node v_1.

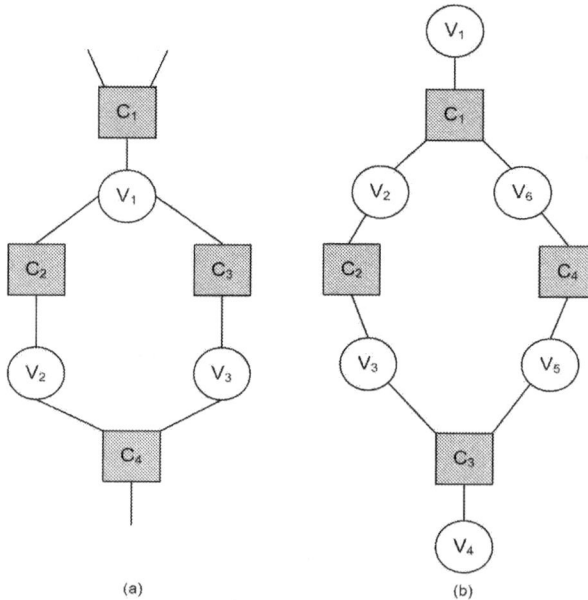

Figure 3.4: Dependency at Variable Node V_1 (a) and Check Node C_1 (b)

Case 2: Dependency between incoming messages to a variable or check node of the graph.

Example: In Fig. 3.4 (b), the optimality of a *check node* is violated. Since check node does not have initial message, the optimality is violated due to the dependency among its incoming messages. The initial message of variable node v_4 will arrive at check node c_1 in second iteration through the paths $(v_4, c_3, v_3, c_2, v_2, c_1)$ and $(v_4, c_3, v_5, c_4, v_6, c_1)$. Hence, the incoming messages from v_2 to c_1 and from v_6 to c_1 are dependent. As a result, in the second iteration the optimality at check node c_1 will be violated and the algorithm is not optimal from the third iteration. It should be noted that iterations start from check nodes.

Based on previous cases, for a *variable* node v in a cycle with length l, the optimality through the cycle will be violated for the first time in iteration number $N = \lceil l/4 \rceil$. For a *check* node c in a cycle with length l, the optimality through the cycle will be violated for the first time in iteration number $N = \lceil (l+1)/4 \rceil$. The proof for these results can be found in [28].

Fact 1: In a Tanner Graph , for a variable or check node u with girth g_u the optimality for the first time is violated in iteration number $N = \lceil g_u/4 \rceil$ for variable node and $N = \lceil (g_u + 1)/4 \rceil$ for check node.

Fact 2: For variable nodes, the message can be returned to the node through a nontrivial closed walk. Therefore, for a variable node u with nontrivial closed walk of length l the local message will be returned to initial node in iteration number $l/2$ for the first time. As a result, if we have a Tanner Graph with cw_u, the optimality of node u will be violated at iteration number $N = cw_u/2$.

For example, in Fig. 3.4 (b) the local message of variable node v_1 will propagate through the closed walk $(v_1, c_1, v_2, c_2, v_3, c_3, v_5, c_4, v_6, c_1, v_1)$ and return back to itself. It happens after 5 iterations. This comes from $\lceil l/2 \rceil$ with knowing the fact that $l = 10$. Therefore, if we have a Tanner Graph with cw_u, the optimality of node u will be violated at iteration number $N = cw_u/2$. Concluding from the facts in this section, I_u is the iteration number in which the optimality of the algorithm is violated for the first time and its value can be calculated based on the following equation:

$$
I_u = \begin{cases} \mathbf{min}\ (\lceil g_u/4 \rceil, cw_u/2) & \textit{If}\ \ u \text{ is a variable node} \\ \lceil (g_u + 1)/4 \rceil & \textit{If}\ \ u \text{ is a check node} \end{cases} \tag{3.1}
$$

3.3 Scheduling

Scheduling is a technique which has great influence on performance and complexity of the decoder. Study on behavior of the iterative decoding algorithms and the graph structure of the code resulted in understanding the Deterministic schedule [28]. This graph-based schedule can be applied on nodes or edges of the graph. In this work, the *node-based unidirectional* and *bidirectional* schedules are applied on Sum-Product and Gallager A decoding algorithms.

3.3.1 Node-Based Schedule

In *Deterministic node-based schedule*, the idea of preserving the optimality is applied on nodes of the graph [28]. The nodes will stop sending messages through the graph based on a pre-determined value. Hence, the name is Deterministic schedule. The pre-determined value shows when the optimality of the node is violated. In previous section, the cases in which the node of a graph may lose its optimality were explained.

After finding the shortest closed walk and girth of the nodes based on the result of a search algorithm, a counter will be assigned to each node with an initial value indicating the iteration number. From this iteration number the algorithm starts losing its optimality. For variable node v, the value of the counter equals to

$$(I_v - 1) = min(\lceil g_v/4 \rceil, cw_v/2) - 1. \tag{3.2}$$

A counter will be assigned to each variable node at the beginning of the iterations. After updating the messages at each iteration, the value of the counter will be reduced by one. If the counter of a node reaches zero , the node will stop sending new messages. From the next iteration, the message of the node has dependency and may cause suboptimality. The node just keeps sending messages from the previous iteration (the last updated message), until the counters for all the variable nodes reach zero. Then we reset the counters and repeat the process till the algorithm converges or maximum number of iterations is reached. For the check nodes the process is the same as variable nodes, but the initial value of the counter is equal to

$$(I_c - 1) = \lceil (g_c + 1)/4 \rceil - 1. \tag{3.3}$$

If the counter is assigned to variable nodes (check nodes), the schedule will be unidirectional from variable nodes (check nodes) to check nodes (variable nodes). If the counter is assigned on both variable and check nodes, the schedule is bidirectional. In the next chapter , the simulation results for unidirectional and bidirectional Deterministic node-based schedules are given.

3.4   Hybrid Decoding

The idea of Hybrid decoding is taken from Gallager B decoding algorithm [2]. In Gallager B algorithm, the outgoing message of a variable node in updating rule are the same as messages coming from the channel, unless at least *"b"* number of the updated messages coming from other check nodes disagree. The value of *"b"* changes from one iteration to another. In other words, Galager B is switching among different decision thresholds during the iterative process. This idea resulted in two techniques called Hybrid Switching-Type (H_{ST}) [35] and Hybrid Time-Invariant (H_{TI}) [29] decoding.

Hybrid (*Multistage*) decoding is using a combination of decoding algorithms and is changing the decoding rule during the iterative process to get improved performance or complexity in the decoder. Hybrid Switch-Type and Hybrid Time-Invariant techniques are explained in this section.

3.4.1   Hybrid Time-Invariant Technique

In Hybrid Time-Invariant technique, N message passing algorithms for decoding will be considered. Specific combination of different algorithms will be used for each iteration. The combination of these algorithms does not change by iteration , hence it is called Time-Invariant technique. In other words, if we consider N message passing algorithms $A_1, A_2, \ldots, A_{N-1}$ for decoding, new algorithm A will be defined based on $A_1, A_2, \ldots, A_{N-1}$. In each iteration of the new algorithm, variable nodes and check nodes are partitioned into N groups using probability mass function vectors of the nodes, $\vec{\alpha}^{(l)} = (\alpha_0^{(l)}, \alpha_1^{(l)}, \ldots, \alpha_{N-1}^{(l)})$ and $\vec{\beta}^{(l)} = (\beta_0^{(l)}, \beta_1^{(l)}, \ldots, \beta_{N-1}^{(l)})$ respectively. The nodes in each group i process the messages with algorithm A_i. The new algorithm A can be shown as :

$$A = H\left(A_1, A_2, \ldots, A_{N-1}, \left\{\vec{\alpha}^{(l)}\right\}_{l=0}^{\infty}, \left\{\vec{\beta}^{(l)}\right\}_{l=0}^{\infty}\right) \qquad (3.4)$$

As an example , consider a regular LDPC code consisting of two algorithms. Variable nodes and check nodes are partitioned into two groups according to their probability mass function vectors $(0.7, 0.3)$ and $(0.5, 0.5)$. Therefore, 70% of the variable nodes and 50% of the check nodes update the messages using the first algorithm. The rest of the variable nodes consisting of 30% of the nodes and the other 50% portion of the check nodes will be updated using the second algorithm [29].

3.4.2 Hybrid Switch-Type Technique

In Hybrid Switch-Type (H_{ST}) method, decoder switches among different decoding algorithms during the iterations. Number of algorithms involved in the decoding process will create the number of stages used in hybrid switching type decoding, hence this technique is also called multistage decoding. In a multistage decoder, the decoding process starts with a few iterations in first algorithm. After a few iterations based on the threshold value switches to the second algorithm and repeats the same trend. This idea can be applied to different algorithms and with different number of stages depending on the application. Based on the number of iterations in different decoding stages and the convergence speed of algorithms very interesting results and complexity/performance trade-offs can be achieved.

The algorithms used for the decoding can have either different or equal complexity. When the complexity of the algorithms are not the same, their computation time and speed of convergence is different, hence choosing a correct combination of algorithms which will result in faster decoding is an important issue. Another issue which should be considered is the compatibility of the algorithms. The transmitted message of different algorithms are not the same, therefore we have to make the output of one algorithm compatible with the input of the other algorithm when transition from one to another occurs.

Since the convergence speed of algorithms are not the same, different number of iterations are required to achieve a message error rate for different decoding rules. Hence, the decoding time for different algorithms is not the same. In this work, decoder starts with a few iterations of soft decision algorithm and then switches to very fast hard decision algorithm, which will result in faster convergence and lower complexity. This technique can not improve the performance of Sum-Product algorithm, because Sum-Product has the best performance among other decoding algorithms but it can reduce the complexity and speed up decoding process. For hard decision algorithms it helps to improve the performance significantly and provide a trade-off between complexity and performance.

In brief, H_{TI} and H_{ST} algorithms are functioning different. In the first method, a combination of algorithms are functioning in each iteration but the ratio of their involvement is not changing in the whole process of decoding. This ratio is variable for different nodes based on their probability mass function. In the second method, all the nodes of one type (variable or check node) use the same algorithm for decoding in each iteration. The decoding algorithm will be changed by switching from one to another during the decoding process in different iterations.

3.5 Search Algorithm to Find Shortest Closed Path

The analysis of LDPC codes to find closed paths can be performed using either by parity-check matrix H or graph of the code G. Finding cycles using H matrix is computationally intensive, therefore we apply graph based search on Tanner Graph of the code. There are many search algorithms to find short cycles or shortest path in the graph, however none of them satisfied our need [41, 48, 49]. In this section, a search algorithm to find the length of the shortest closed-walk and shortest cycle of the nodes is introduced. The results are used to initialize the iteration numbers in Eqs. 3.2 and 3.3.

Fig. 3.5 illustrates part of the graph structure of a specific code. At the beginning, we assume that graph has a tree structure. In this algorithm, a series of interconnected nodes will be searched through to find the shortest closed path for each node. A node can be connected to other nodes via edge or path. The nodes which are connected to each other are called neighbors. The search can be started from any desired node in the graph. This node is called root node (start node). The graph will be traversed until the search results in a closed path (cycle or closed-walk).

3.5.1 Graph-Based Search Algorithm

The idea of the proposed algorithm is to find the length of the shortest closed path for each node in the Tanner Graph of the code. Let's assume that a closed path is encountered at layer m of the graph. It can be concluded that the graph will have structure of a tree up to the layer $m - 1$, otherwise this would not be the first closed path we come across in our search. There are usually different edges coming out of each node in the tree. Each of these edges create a branch initiated from the start

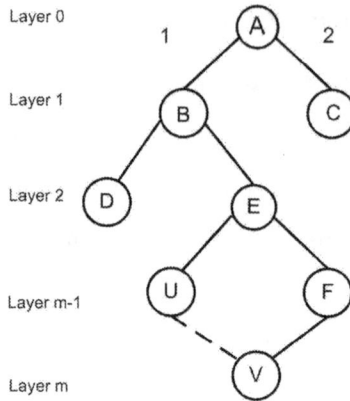

Figure 3.5: Tree Structure of the Code

node. The number of edges is equivalent to the number of connections for the node in the parity-check matrix. The outgoing edges of a node can be labeled using sequential numbers $1, 2, 3, \ldots$. If at one layer, node v connects with two nodes from previous layer, this will result in a closed path from the start node to node v.

The closed path for node v can be a cycle or a closed walk. Closed path will be a cycle, if the connections from previous steps for node v are initiated from two different edges of the root node (start node), otherwise it is a closed walk. The length of the path is twice the number of layers from the start node to the end node. Different steps of the search algorithm are outlined as follows:

for each vertex $u \in V[G]$ do:

1: **Initialization:**

$r = u$, $flag = false$, $length = 0$, $cycle = false$, $cw = false$

2: **begin**

3: **while**$(flag = false)$

4: layer := layer+1

5:      **if** $n(v \in Ajc\,[v_n]) \succ 2$ ($n$is the number of vertices adjacent to node v_n)

6:          **then** ($flag = true$)

7:      **if** $u \in path(v_1, v_2)$

8:          **then** ($cycle = true$)

9:              **else**

10:                 ($cw = true$)

11:                 $length = 2\times$ layer

12: **end**

Based upon the above descriptions, we will now present a lemma and a theorem to prove the fact that the algorithm will find the shortest cycle and shortest closed walk passing through each node of the graph G.

Corollary 1: A graph is a tree, if and only if, for every pair of distinct vertices u and v there is only one (u, v) path.

Let's assume node A is our root node. If node A is connected to k other nodes, there will be exactly k edges coming out of root node A.

Lemma: If in any layer of a graph, one node is connected to at least two nodes from the previous layer, a closed path is formed and the graph will not be a tree anymore.

Proof: Suppose G is a tree and (u, v) is an edge not included in G. Let's assume u is in layer $m-1$ and v is in layer m. By corollary 1, there is exactly one (u, v) path in the graph which consists of a connection from layer m to layer $m-1$. This will be the first connection from node v in layer m and layer $m-1$. If the path between node u and v is $ux_1x_2\ldots x_kv$, then adding edge (u, v) which is the second edge between these two layers consisting node v, will create the cycle $ux_1x_2\ldots x_kvu$ and prove the lemma.

Based on the definition of a tree there is a unique path between any nodes of the cycle and node A. If the cycle involves node A, we have found a cycle going through root node. If the cycle does not involve node A, a closed walk is found which is going though root node. By unfolding the tree from the root node (start node), there are usually different branches coming out of each node in the tree. The number of branches is equivalent to the number of connections or degrees of the node in the parity-check matrix. These lead us to the following theorem.

Theorem: Suppose node A is the root node in graph G with 2 branches which are labeled as 1 and 2. If node v in layer m of the graph is connected to two nodes from different branches, it creates a cycle that will go through root node A. If node v in layer m of the graph is connected to two nodes from same branch, the cycle will not include node A. In this case, we have found a closed walk which is going through node A.

Proof: Any node that is part of a cycle should have at least two branches. Therefore, root node A can be part of a cycle if its degree is at least 2. As a result we conclude that if A is part of a cycle, a node in layer m has to be connected to two nodes from different branches in layer $m-1$. Furthermore, let's assume that the cycle does not involve node A. This means that there is a smaller cycle between layer m and layer n ($0 \prec n \prec m$). If this is the case, then there will be more than one path between a node in layer n and layer $n+1$ which does not include edges from layer $m-1$ to m. This will be in contradiction to the assumption of a tree structure up to the layer m and before finding the first closed path in the graph. Therefore, based on the theorem we have found a closed walk which is going through node A.

3.5.2 Illustrative Example

The following example is designed to better illustrate the application of the algorithm. In Fig. 3.6, parity-check matrix *H1* with 9 variable nodes and 6 check nodes is illustrated. The lines on the matrix are showing the patterns of the cycles with length 4 and length 6. Variable nodes $\{v_3, v_5\}$ and check nodes $\{c_2, c_4\}$ are involved in the cycle with length 4. Variable nodes $\{v_0, v_4, v_6\}$ and check nodes $\{c_0, c_1, c_5\}$ are involved in the cycle with length 6.

$$
H1 = \begin{array}{c c} & \begin{array}{ccccccccc} v_0 & v_1 & v_2 & v_3 & v_4 & v_5 & v_6 & v_7 & v_8 \end{array} \\ \begin{array}{c} c_0 \\ c_1 \\ c_2 \\ c_3 \\ c_4 \\ c_5 \end{array} & \left[\begin{array}{ccccccccc} 1 & 0 & 1 & 0 & 1 & 0 & 0 & 0 & 1 \\ 0 & 1 & 1 & 0 & 1 & 1 & 1 & 0 & 0 \\ 0 & 0 & 0 & 1 & 0 & 1 & 0 & 0 & 0 \\ 0 & 0 & 0 & 1 & 1 & 0 & 1 & 1 & 0 \\ 0 & 1 & 1 & 1 & 0 & 1 & 0 & 0 & 1 \\ 1 & 1 & 0 & 0 & 0 & 0 & 0 & 1 & 0 \end{array} \right] \end{array}
$$

Figure 3.6: Parity-Check Matrix with 4-Cycle and 6-Cycle

$$
H2 = \begin{array}{c c} & \begin{array}{ccccccccc} v_0 & v_1 & v_2 & v_3 & v_4 & v_5 & v_6 & v_7 & v_8 \end{array} \\ \begin{array}{c} c_0 \\ c_1 \\ c_2 \\ c_3 \\ c_4 \\ c_5 \end{array} & \left[\begin{array}{ccccccccc} 1 & 0 & 2 & 0 & 2 & 0 & 0 & 0 & 2 \\ 0 & 0 & 0 & 0 & 3 & 0 & 3 & 0 & 0 \\ 0 & 0 & 0 & 0 & 0 & 0 & 0 & 0 & 0 \\ 0 & 0 & 0 & 0 & 3 & 0 & 3 & 0 & 0 \\ 0 & 0 & 0 & 0 & 0 & 0 & 0 & 0 & 0 \\ 1 & 2 & 0 & 0 & 0 & 0 & 2 & 2 & 0 \end{array} \right] \end{array}
$$

Figure 3.7: Closed Path with Length 6 for Variable Node v_0

$$
H3 \quad = \quad
\begin{array}{c c}
& \begin{array}{c c c c c c c c c} v_0 & v_1 & v_2 & v_3 & v_4 & v_5 & v_6 & v_7 & v_8 \end{array} \\
\begin{array}{c} c_0 \\ c_1 \\ c_2 \\ c_3 \\ c_4 \\ c_5 \end{array} &
\left[
\begin{array}{c c c c c c c c c}
1 & 0 & 1 & 0 & 1 & 0 & 0 & 0 & 1 \\
0 & 0 & 0 & 0 & \boxed{1} & 0 & \boxed{2} & 0 & 0 \\
0 & 0 & 0 & 0 & 0 & 0 & 0 & 0 & 0 \\
0 & 0 & 0 & 0 & 1 & 0 & 2 & 0 & 0 \\
0 & 0 & 0 & 0 & 0 & 0 & 0 & 0 & 0 \\
2 & 2 & 0 & 0 & 0 & 0 & 2 & 2 & 0
\end{array}
\right]
\end{array}
$$

Figure 3.8: Cycle for Variable Node v_0

To find the shortest cycle and closed walk passing through variable node v_0 which has length 6 in this example, the search is started in the graph of the code which is adjacent to the parity-check matrix. Unfolding the tree will be started from root node v_0. Since v_0 is connected to 2 check nodes c_0 and c_5, its degree is equal to 2 and two branches at the first step are labeled by 1 in $H2$ matrix (Fig. 3.7) to indicate the connection between the variable and the check nodes. Two different branches are labeled as 1 and 2 in $H3$ matrix to indicate two different paths (Fig. 3.8).

The traverse of the tree is continued in the second layer by finding the connections between check nodes $\{c_0, c_5\}$ and unvisited variable nodes. The connections are labeled as 2 in $H2$ matrix . The possibility of having a closed path is checked in each layer of the graph. In matrix $H3$, the connections are labeled as 1 and 2 for the nodes initiated from branches 1 and 2 respectively.

In layer 3, the connections between variable nodes $\{v_1, v_2, v_4, v_6, v_7, v_8\}$ and the check nodes in successive layer are detected. The connections for $\{v_4, v_6\}$ are labeled as 3 in matrix $H2$. Based on the lemma, since check node c_1 is connected to two variable nodes $\{v_4, v_6\}$ from previous layer, a closed path is found for root node v_0. Furthermore, as it is shown in $H3$ matrix, these variable nodes are initiated from

46

different branches. Therefore, based on the theorem the closed path is recognized as a cycle for root node v_0. The length of the closed path is two times by the number of steps or layers traversed from root node v_0 to c_1 which is equal to 6. The search will continue through the graph, until we reach to the first closed walk for the root node. For the sake of simplicity, the connections of other nodes are not marked in matrices $H2$ and $H3$.

3.6   Multistage Scheduled Decoder

It has been shown that Sum-Product (SP) algorithm has the best performance among all message-passing algorithms but it has very high complexity [16]. Gallager A (GA) algorithm is much less complex than SP and is suitable for high throughput systems and implementations, but it does not provide desirable performance. Therefore, we are looking for a decoder with desirable performance/complexity trade-off. In this work, we apply the node-based Deterministic schedule on each algorithm to preserve the optimality and improve their performance individually. After a few iterations of SP, the decoder switches into a less complex algorithm GA for updating the messages. This technique offers a good trade-off between decoding complexity and error performance by combining a simple hard-decision algorithm and a soft-decision algorithm with proper scheduling.

3.6.1   Decoding Strategy

Before starting the decoding algorithm, the length of the shortest cycle and closed walk of each variable node and check node is derived using some search programs. Girth and closed walk are used for initializing Eq. 3.2 and Eq. 3.3.

Fig. 3.9, shows an overview of the decoding process. In initialization step, the

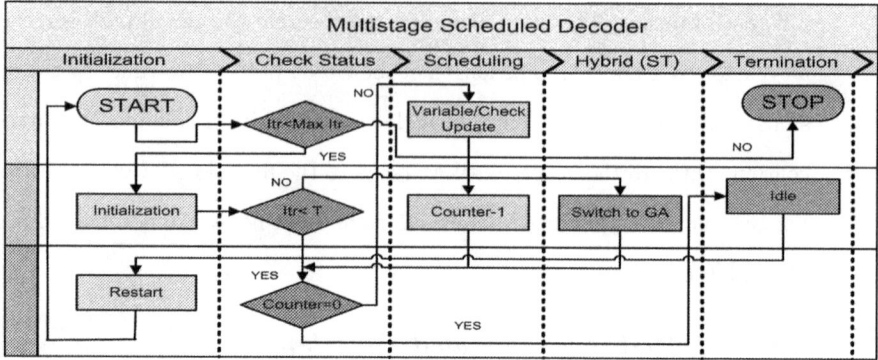

Figure 3.9: Flowchart of The Decoding Process

counters will be initialized and assigned to check nodes and variable nodes. The threshold value (T), which determines the transition time to the new stage will be initialized too. Based on these values, some nodes may always take part in update process while others stop updating after a few iterations as soon as losing their optimality. The latter group of nodes will stay in idle state until all the nodes in the same set lose their optimality. Then, they restart updating the messages with their initial values. If the threshold is not reached, the algorithm starts scheduling. The decoder consists of two stages. At the first stage, a Soft-decision algorithm (Sum-Product) will start updating the messages. This step is shown as START point in Fig. 3.9. Deterministic schedule can be applied on variable nodes, check nodes or both sets. If number of iterations reach to its maximum value, the algorithm will go to the STOP point. After a few iterations, the optimality of the algorithm is preserved and improvement in the performance is obtained partially. Therefore, weak performance of GA is compensated and we can switch to the less complex hard-decision algorithm. This transition, from first stage to second stage, occurs after a few iterations which

48

is indicated by threshold value.

The threshold value can be set based on desirable complexity/performance trade-off and will be checked after each updating rule. In this work, we apply Hybrid Switch-Type (ST) decoding on our algorithms. Therefore, we switch to a different decoding algorithm after T number of iterations. We make the output of one algorithm compatible with the input of the other. The transition occurs at the output of the variable nodes in SP and converts soft information to binary messages:

$$
\begin{cases}
m = \frac{1-sign(m)}{2} & \textit{If } |m| > 0 \\
m = 0, 1 \ \textit{randomly} & \textit{If } |m| = 0
\end{cases}
\tag{3.5}
$$

Same as previous stage, the algorithm may function conventionally (Flooding schedule) or update the messages using the node-based schedule. The decoder stops, if the codeword is found or maximum number of iterations is reached.

In this work, the idea is applied on short block length LDPC codes due to the loopy nature of their graphs with many short cycles and the importance of preserving the optimality at their decoder side. This issue has less significance for large block length codes due to the long cycles in their TG structure, where dependency disappears by propagating in the graph after some iterations. The schedule is applied on check nodes, variable nodes and check/variable nodes of SP and GA algorithms and the threshold value is set to $T = 2, 3, 5$.

3.6.2 Advantages of the New Decoder

- The idea of combining Deterministic schedule and Hybrid Switch-Type technique can be applied on any iterative decoding algorithm and different available

LDPC codes.

- The decoding algorithm does not deal with probability calculations or random generators, which results in less complexity compared to conventional decoding schemes.

- The flexibility of the decoder allows to obtain desirable results by simply changing predetermined values.

- After applying the schedule, propagation of unreliable information in the Tanner Graph of the code is prevented.

- Hybrid technique provides an option for choosing the decoding rule among different decoding algorithms based on desirable performance/complexity trade-off.

- In contrast with conventional Flooding schedule, all the nodes are not taking part in update process at each iteration. Hence, the complexity at each iteration and as a result in the algorithm is reduced.

3.7 *Summary*

In this chapter, the main theme of this thesis was defined and the advantages of the work were indicated. In section 3.1, some necessary background on Tanner graph structure of the code was provided and some definitions on graphs were given. In the second section, the suboptimality cases for decoding algorithms were studied and 2 suboptimality cases were indicated.

These studies helped us to understand Deterministic node-based schedule, which is a graph-based schedule. This schedule was explained in the third section. The schedule finds the iteration number in which a node in the graph loses its optimality. Then, stops the node from updating messages and preserves the sub-optimality of the decoding algorithm. In section 3.4, the Hybrid Switch-Type and Hybrid Time-Invariant techniques were studied.

Based on the studies in section 1 and 2, a search algorithm to find the length of the shortest closed walk and shortest cycle for each node was developed. This algorithm was provided in section 3.5. Also, a theorem and a lemma used to design the search algorithms were presented and proved. In section 3.6, the main idea of the work which was combining the Deterministic schedule and Hybrid Switch-Type technique was described in detail. Moreover, the advantages of this decoder were outlined.

CHAPTER 4

SIMULATION RESULTS

In this chapter, the simulation system model of the work is described and simulation results for studying the performance of multistage decoder are provided. The simulation results are based on different codes and iterative decoding algorithms. Furthermore, decoding complexity and statistics of iteration numbers for different algorithms are analyzed and discussed.

The organization of this chapter is as follows. In section 4.1, the model of our communication system is depicted and parity-check matrix generation is described. In section 4.2, system model of the work is explained. In section 4.3, the simulation results and performance discussion for a regular (1200,600) and irregular (1008, 504) codes are provided. Section 4.4 provides the result of a comprehensive study on complexity of the algorithms for both regular and irregular codes. The summary of this chapter is given in section 4.5.

4.1 Parity-Check Matrix Generation

The simulation system model of the work is illustrated in Fig. 4.1. First, sparse matrix generator provides the sparse parity-check matrix of the code "H". The first code which is generated in this work is a $(n, k) = (1200, 600)$ regular LDPC code with rate $1/2$, block length of 1200 and number of information bits 600 (dimension of the code). The code is constructed randomly with weight of the columns $w_c = 3$ and weight of the rows $w_r = 6$ using Mackay construction method "1A" [8]. The

52

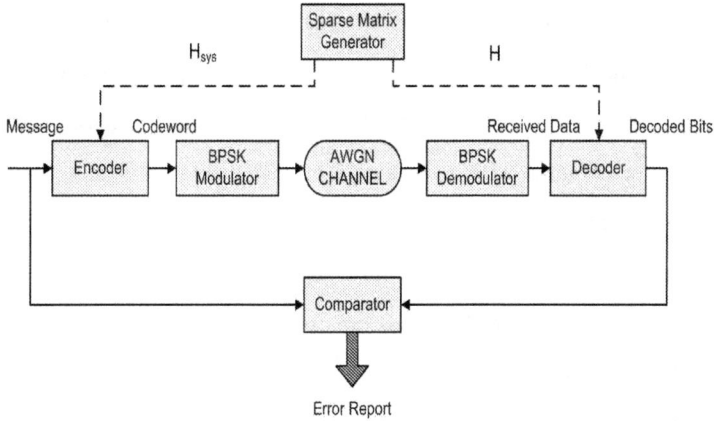

Figure 4.1: Block Diagram of Simulation System Model

experiment is repeated with an irregular LDPC code (1008, 504) with rate 1/2, degree of the columns $\{2, 3, 4, 5, 7, 14, 15\}$ and degree of the rows $\{7, 8, 9\}$ which is optimized for AWGN channels. Both codes are available in [50]. For the encoding process, a generator matrix dual to the parity-check matrix is generated. This matrix is systematic "H_{sys}", therefore the message bits appear in the codeword and make the decoding process easier. In order to reduce the effect of very short cycles and increase the performance of the code, all the cycles with length 4 are removed from "H" after the generation.

All the graphs in this work are Bipartite, therefore any closed path with no repeated node (cycle) in the graph of the code must have an even length. The graph of the code contains many cycles with different lengths, however we focus on the shortest possible cycle which has length of 4. An algorithm is applied on adjacency matrix of the graph "H" to locate and remove 4-cycles.

Any cycle with length 4 involves four distinct nodes and four distinct edges. Consequently in the adjacency matrix four 1s in unique positions create the cycle [48, 51]. After designing the parity-check matrix H, the algorithm searches the parity-check matrix for columns with two 1s in identical positions forming rectangle of four 1s in the matrix. The algorithm tries to eliminate the rectangle, while preserving the properties of the matrix. The most important concern in removing 4-cycles is to make sure that no new 4-cycle is created and all the characteristics of the code are preserved. This can be achieved by swapping edges (1s) in a way that no new cycle is formed. By applying the algorithm on any LDPC code, a new code with equal size and complexity will be created which results in faster and more accurate decoding.

4.2 System Model

After the generation of H_{sys} and H through sparse matrix generator, the system can start its functionality. Messages are generated randomly which consist of 0s and 1s. After encoding, data is modulated using Binary Phase shift Keying (BPSK) modulation technique. As a result -1 for a 0 bit and $+1$ for a 1 bit are sent through the channel. The received data from the channel consists of the codeword and added random noise and is sent to the decoder after demodulation process. Decoder proceeds in different stages and passes the messages along the edges of the graph. It starts with the initialization stage, alternates between check node and variable node update stages and stops at a terminal stage. Variable nodes receive intrinsic messages from channel and extrinsic updated messages from check nodes. The new updated messages from variable nodes will be sent to check nodes in the next half iteration. The maximum number of iterations is set to 50. If the equation $\hat{c}H^T = 0$ is satisfied, the codeword c is found at the output of the decoder and the algorithm stops decoding. Otherwise, the iteration will continue to update the messages until the maximum number of iterations is reached. If the decoder fails after reaching the maximum number of iterations, another new message will be generated and passed through the encoder, modulator, AWGN channel, demodulator and decoder. At each E_b/N_o enough codewords are created to generate 100 codeword errors, hence procedure continues until we find 100 errors. The comparator will compare the output of the decoder to the input of the encoder, in order to compute bit error rate and message error rate.

4.3   Performance Study of the Decoding Algorithms

In this section, simulation results are provided to investigate the performance of multistage scheduled decoder for regular and irregular LDPC codes. In chapter 3, a Deterministic node-based schedule was applied on the nodes of the graph. The schedule was indicating the iteration number in which the node loses its optimality. Therefore, the node was stopped updating messages and the optimality of the algorithm was preserved. Furthermore, Hybrid Switch-Type technique with different threshold was applied on the improved decoding algorithms to reduce the complexity and provide a desirable performance/complexity trade-off. In order to study the behavior of the decoder, the idea is applied on soft-decision algorithm SP and hard-decision algorithm GA. The simulation results show that in all cases, better performance can be achieved compared to conventional algorithms except SP. Also, the performance of irregular LDPC code is considerably better than regular LDPC code. The bit error rate (BER) curves are represented with solid lines and the message error rate (MER) curves are represented with dashed lines in all Figures.

4.3.1 Performance Study of (1200, 600) Regular Code

In Fig 4.2 and Fig 4.3 the effect of applying Deterministic schedule on nodes of the graph is presented. These results prove that, the schedule applied on the nodes helps to preserve the optimality of the algorithms.

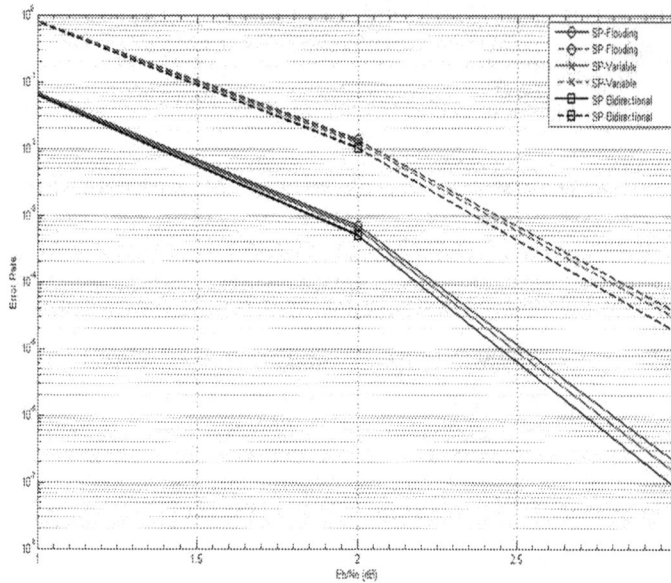

Figure 4.2: BER(–) and MER(- -) for SP Algorithm with (1200, 600) Code

The regular code which is used for performance study in this work is a $(n, k) = (1200, 600)$ randomly constructed LDPC code with rate $1/2$, weight of the columns $w_c = 3$ and weight of the rows $w_r = 6$. Fig. 4.2 shows the bit error rate and message error rate curves for SP algorithm with Flooding schedule, SP with Deterministic schedule applied on variable nodes (SP-Variable) and SP algorithm with

Deterministic schedule applied on variable nodes and check nodes of the algorithm (SP-Bidirectional).

It can be observed that, SP-Variable algorithm has better performance compared to conventional (SP-Flooding) algorithm. This is due to the fact that, unreliable variable nodes do not take part in update process and the reliability of the messages are increased in this algorithm. Based on the same reason, SP-Bidirectional has better performance in comparison with other two algorithms.

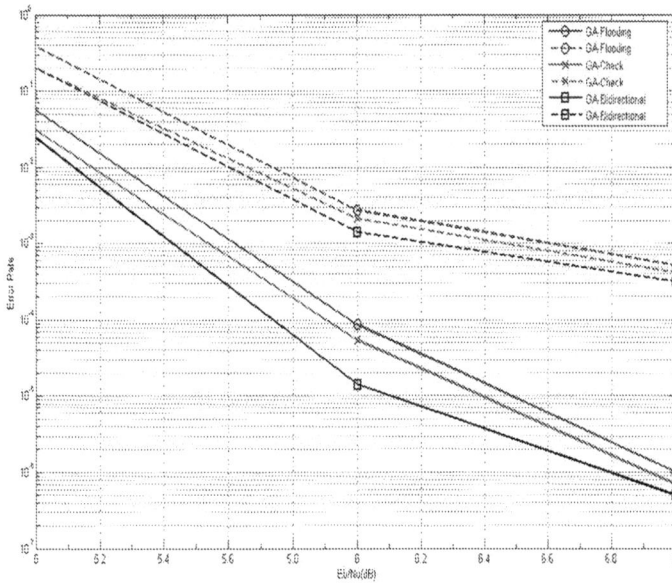

Figure 4.3: BER(–) and MER(- -) for GA Algorithm with (1200, 600) Code

Fig. 4.3 shows the bit error rate and message error rate curves for GA algorithm with Flooding schedule (GA-Flooding), GA with Deterministic schedule applied on check nodes (GA-Check) and GA algorithm with Deterministic schedule applied on variable nodes and check nodes of the algorithm (GA-Bidirectional). The same as SP algorithm, performance of the GA algorithms with applied schedules are better than conventional Flooding schedule (GA-Flooding). However, based on TG structure of the randomly generated code and its cycle distribution different performance results can be obtained.

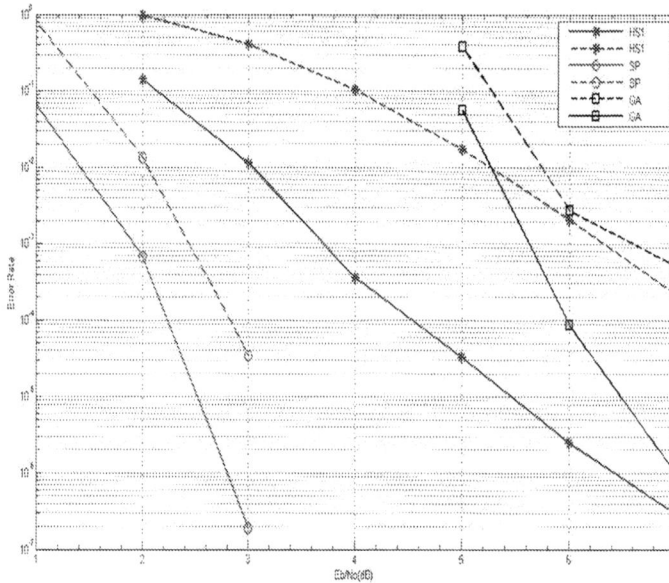

Figure 4.4: BER(–) and MER(- -) for HS1 Algorithm with (1200, 600) Code

In Figs. 4.4-4.6, the bit error rate and message error rate of the new algorithms are plotted. The simulation results for GA and SP algorithms with Flooding schedule are provided for the reference. In Fig 4.4, the result of first experiment is shown as curve HS1.

The Deterministic node-based schedule is applied to variable nodes of SP algorithm and the threshold value is set to $T = 2$, therefore the decoding starts with SP algorithm and after 2 iterations it switches from SP to GA. The check nodes of SP algorithm and GA algorithm update the messages in conventional way using Flooding schedule. It can be observed that for BER= 10^{-6} almost 0.6 dB gain is obtained compared to GA algorithm, but SP algorithm still has the best performance.

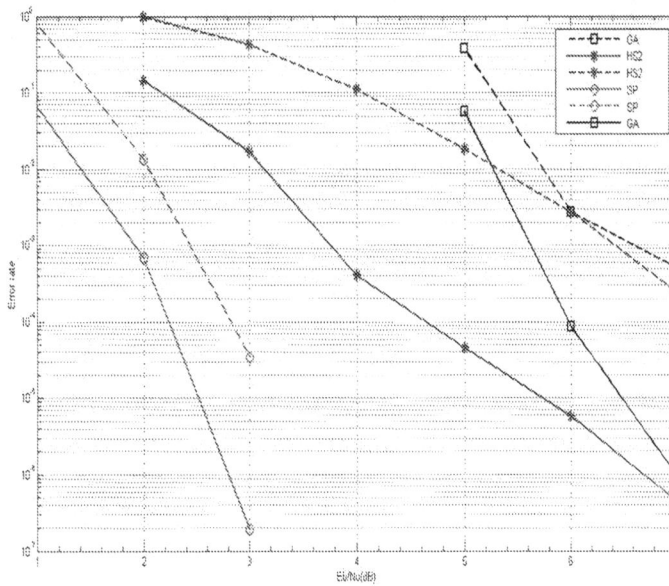

Figure 4.5: BER($-$) and MER(- -) for HS2 Algorithm with (1200, 600) Code

In Fig. 4.5, the simulation results for HS2 algorithm is provided. The schedule is applied on variable nodes of SP and GA algorithms and the threshold value is set to

$T = 2$. At this experiment, the only difference from HS1 algorithm is application of the schedule on variable nodes of the GA algorithm. Therefore, it can be observed that there is no considerable improvement in the performance. In fact, since the code is constructed randomly, the error performance is dependent to the distribution of the cycles in the graph of the code which results in no improvement in this case compared to HS1 algorithm.

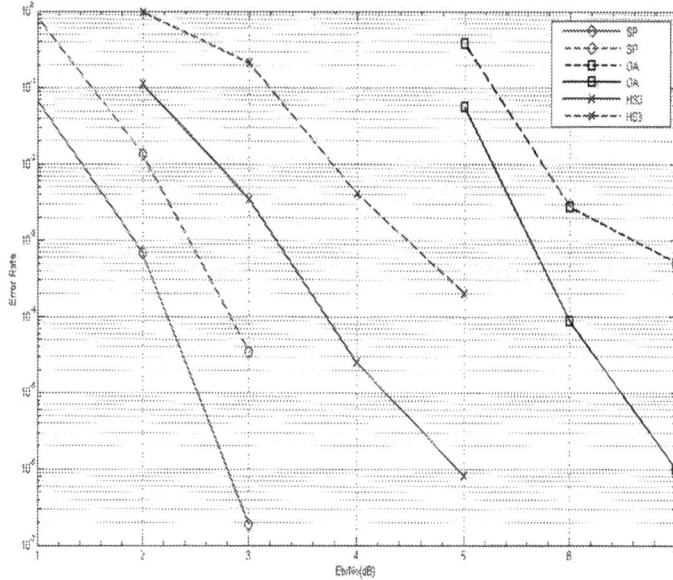

Figure 4.6: BER(–) and MER(- -) for HS3 Algorithm with (1200, 600) Code

For the third experiment, the threshold value is changed from $T = 2$ to $T = 5$ and Deterministic node-based schedule is applied on variable nodes of SP algorithm and check nodes of GA. The decoder starts with SP and switches to GA after 5 iterations.

It can be observed from HS3 in Fig. 4.6 that the improvement in the performance is considerably more that HS1 and HS2. This observation can be explained in the way that decoder is spending more time (5 iterations) in SP algorithm, hence more reliable information is transfered to the next stage of the decoder. It can be observed that for BER= 10^{-6}almost 2 dB gain is obtained compared to GA algorithm.

4.3.2 Performance Study of (1008, 504) Irregular Code

To investigate the performance of the decoder with irregular LDPC codes, the experiment is repeated with an irregular LDPC code (1008, 504) with rate 1/2, degree of the columns $\{2, 3, 4, 5, 7, 14, 15\}$ and degree of the rows $\{7, 8, 9\}$.

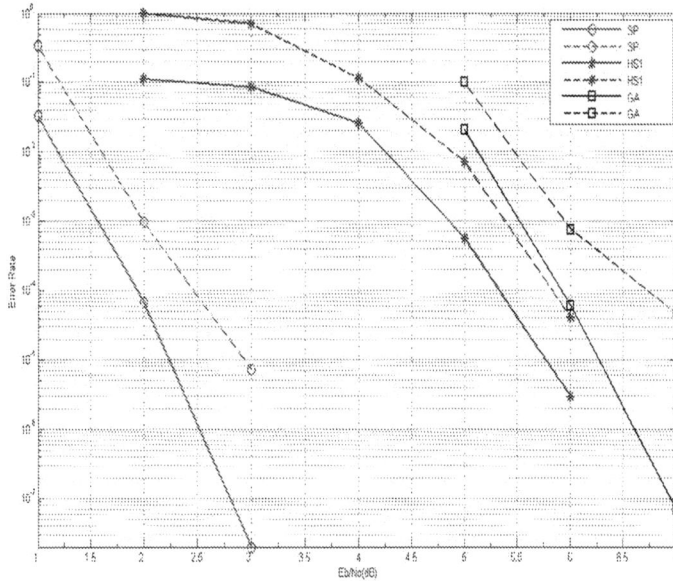

Figure 4.7: BER(–) and MER(- -) for HS1 Algorithm with(1008, 504) Code

61

Due to the nature of irregular codes, they show better performance results compared to regular (1200, 600) code [38]. Same as the previous trends, the idea is applied on SP and GA algorithms. In Fig. 4.7, the Deterministic schedule is applied on variable nodes of the SP algorithm. The check nodes of SP and variable/check nodes of GA algorithms update messages using Flooding schedule. The threshold value is set to $T = 2$, hence after 2 iterations the decoder switches from SP to GA algorithm.

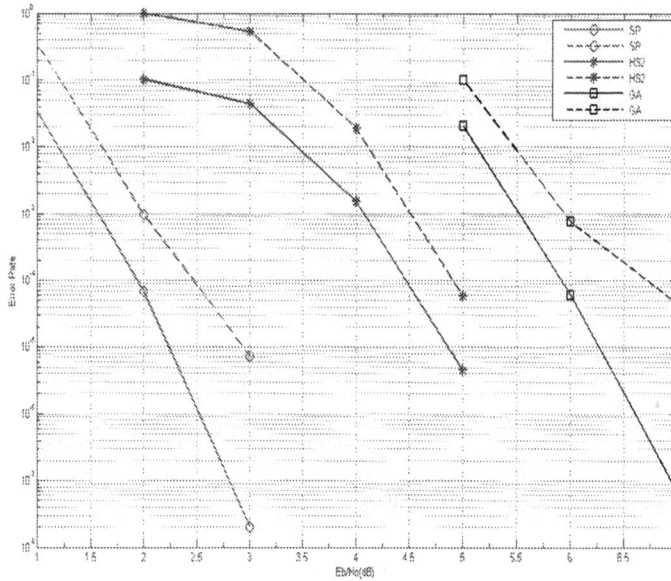

Figure 4.8: BER(–) and MER(- -) for HS2 Algorithm with (1008, 504) Code

Fig. 4.8 shows the performance of the decoder when schedule is applied on variable nodes of SP algorithm and variable nodes of GA. The threshold value is set to $T = 3$.

It can be observed that the algorithm has considerable improvement compared to GA algorithm and is getting closer to SP. Also, It has better performance compared to HS1 algorithm with irregular code due to the fact that more reliable information will transfer to GA algorithm after 3 iterations. We can also observe that at BER= 10^{-5} HS2 algorithm with irregular code has obtained almost 1 dB gain compared to HS2 algorithm with regular code. Fig 4.9, shows the performance result of HS3 algorithm with threshold of $T = 5$. The schedule is applied on variable nodes of SP algorithm and check nodes of GA algorithm. The algorithm shows very good performance because of a high threshold value and irregularity of LDPC code.

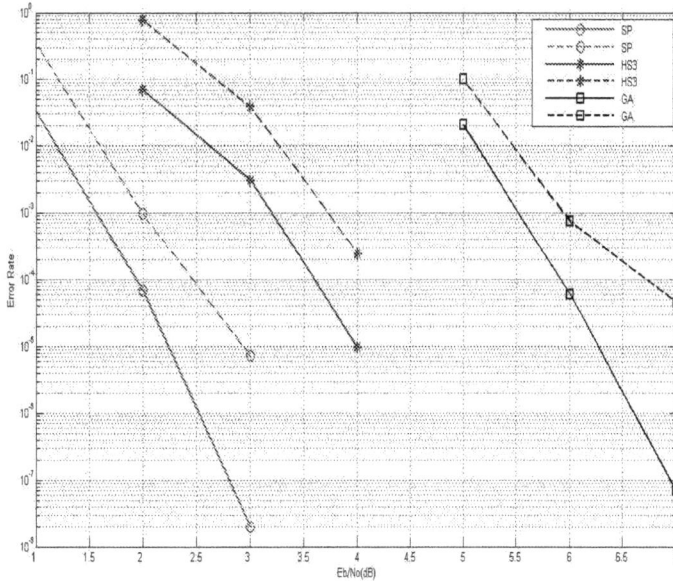

Figure 4.9: BER($-$) and MER(- -) for HS3 Algorithm with (1008, 504) Code

4.4 Complexity Study of the Decoding Algorithms

In LDPC codes, which are graph-based, the decoding complexity depends on the number of edges E in the Tanner Graph of the code. The number of edges is equal to number of ones in the parity-check matrix of the code. Decoding complexity also depends on the number of iterations. As described earlier, decoding terminates if the codeword is found or after maximum number of iteration when the syndrome decoding fails. In this section we study the complexity of algorithms based on the above factors [52].

4.4.1 Complexity Study of (1200, 600) Regular Code

Table 4.1 provides the statistics on complexity of decoding for regular (1200, 600) code.

Table 4.1: Study of the Complexity of Algorithms for (1200, 600) Code

Algorithms	Schedule	Involved Nodes	Eb/No(dB)	Average No. of Soft Iterations	Average No. of Hard Iterations	Complexity
SP	Flooding	------	1 2 3	44.6 10.4 5.1	0 0 0	Very High
GA	Flooding	------	5 6 7	0 0 0	25.1 4.8 2.3	Very Low
HS1	Deterministic	Variable/------	4 5 6	2 2 2	7.2 1.3 0.1	Low
HS2	Deterministic	Variable/Variable	4 5 6	2 2 2	7.4 1.3 1.2	Low
HS3	Deterministic	Variable/Check	2 3 4	5 5 5	43.7 9.7 0.1	Medium

Table 4.1 provides the average number of soft/hard iterations required for convergence of algorithms at their waterfall region for regular (1200, 600) code. Due to the fact that Sum-Product algorithm has a lot of computations in its variable/check node update rule, it has highest decoding complexity among other decoding algorithms. GA has lowest complexity among others, since it does not deal with soft information and its average number of soft iterations is 0. Based on these analysis, SP provides the desirable performance at lower $Eb/N0$ compared to other algorithms with the cost of complexity. In HS2, we can achieve to the desired performance at lower $Eb/N0$ compared to GA by performing 2 iterations of SP followed by required number of hard iterations in GA. GA provides lowest complexity at the cost of performance loss.

Table 4.2: Study of the Complexity at Bit Level for (1200, 600) Code

Algorithms	Eb/No(dB)	BC_{AVE}	BC_{MAX}	Number of Edges
SP	1	298.64	301.1	3614
	2	63		
	3	30.8		
GA	5	175.6	300.8	3610
	6	25.27		
	7	13.8		
HS1	4	55.3	300.5	3607
	5	19.8		
	6	12.6		
HS2	4	56.7	301.3	3616
	5	20.6		
	6	2.5		
HS3	2	293.4	301.3	3616
	3	88.59		
	4	30.7		

Therefore, this table presents the complexity/performance trade-off that is provided in this work. It can be observed that increasing the complexity (number of soft iterations) results in better performance (lower $Eb/N0$).

In table 4.2, the complexity per bit for different algorithms are provided and compared in their waterfall region. BC_{AVE} is the per bit complexity based on average number of iterations "I_{AVE}" and BC_{MAX} is the per bit complexity based on maximum number of iterations "I_{MAX}". They can be calculated using following equations,

$$BC_{AVE} = \frac{I_{AVE} \times E}{K} \tag{4.1}$$

$$BC_{MAX} = \frac{I_{MAX} \times E}{K} \tag{4.2}$$

where E is the number of edges in the graph and K is the number of information bits.

The first equation helps us design a decoder for average number of iterations and choose appropriate I_{AVE}, in order to achieve desirable performance. For example, algorithm SP at $Eb/No = 3$ dB has $BC_{AVE} = 30.8$, while $BC_{MAX} = 301.1$ is almost 10 times larger. Based on the results on this table, we can conclude that I_{AVE} is significantly smaller than I_{MAX} in our algorithm. Therefore, we can design decoders with lower maximum number of iterations which results in complexity reduction.

Figs. 4.10-4.14 give the statistical analysis of each iteration number for different values of Eb/No in waterfall region of the algorithms. The percent on utilization level (UL%) of each iteration number helps us to study the behavior of the algorithm. For small $Eb/N0$, most decodings fail and required number of iterations is almost equal to I_{MAX}. In higher Eb/No, smaller iterations become dominant which results in lower I_{AVE}. It can be observed that at higher Eb/No, I_{AVE} is much less than I_{MAX}.

As an illustrative example, we consider Fig. 4.12 for HS1 algorithm. The required number of iterations for $Eb/No = 2.0$ dB is almost equal to 50 (more than 90%). Therefore, we can conclude that for this algorithm at $Eb/No < 3$ dB the decoding requires maximum number of iterations. By increasing Eb/No, the number of smaller iterations increases which results in lower average number of iterations. Therefore, the significance of I_{MAX} is reduced in higher Eb/No (Fig. 4.12 (d)). The significant difference between I_{AVE} and I_{MAX} for high Eb/No has a great influence on complexity of the algorithm as it was shown in table 4.2.

The statistical study of the complexity for LDPC codes has a significant role in designing a low complex decoder. One of the important parameters in the design is number of decoding iterations which varies based on Eb/No values. PDFs of the iteration numbers give sufficient information on decoding complexity which are used for implementation issue. These results are given for all the algorithms in this work and are used for evaluation of trade-off between performance of LDPC codes and their complexity.

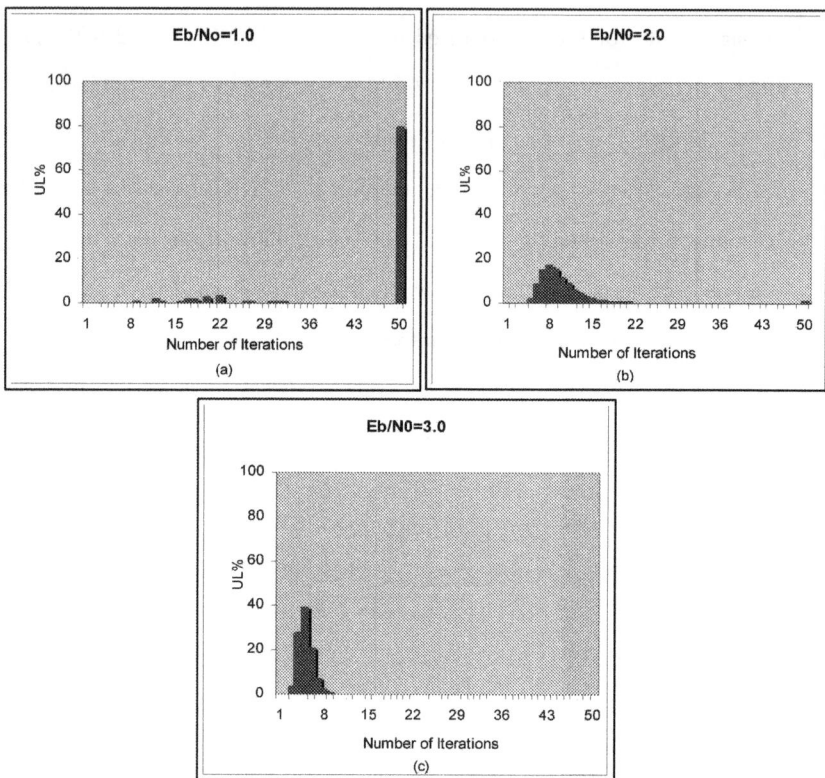

Figure 4.10: *Pdfs* of the Iterations for SP Algorithm with (1200, 600) Code

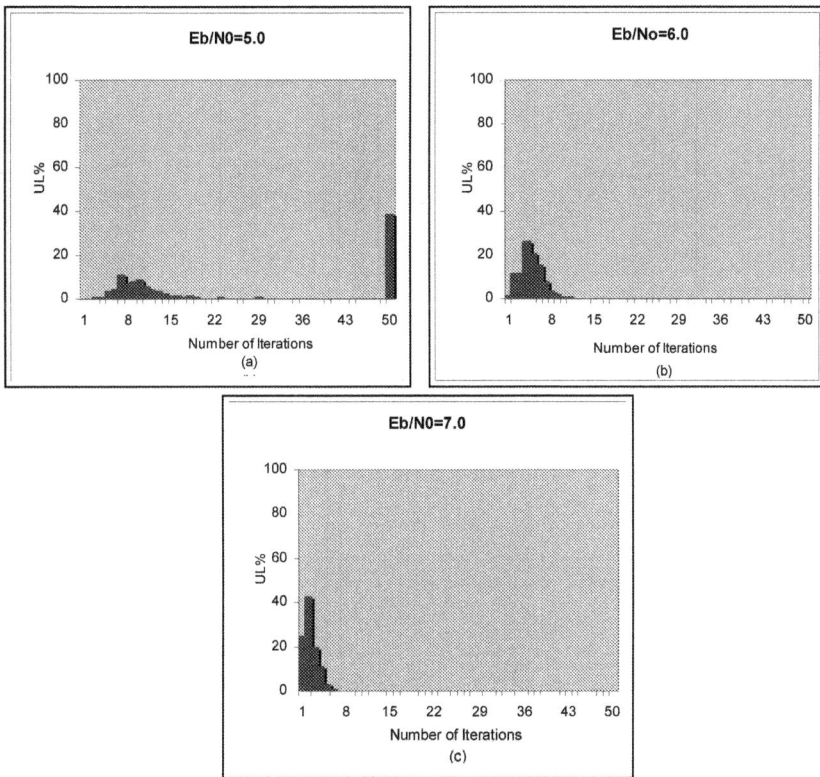

Figure 4.11: *Pdfs* of the Iterations for GA Algorithm with (1200, 600) Code

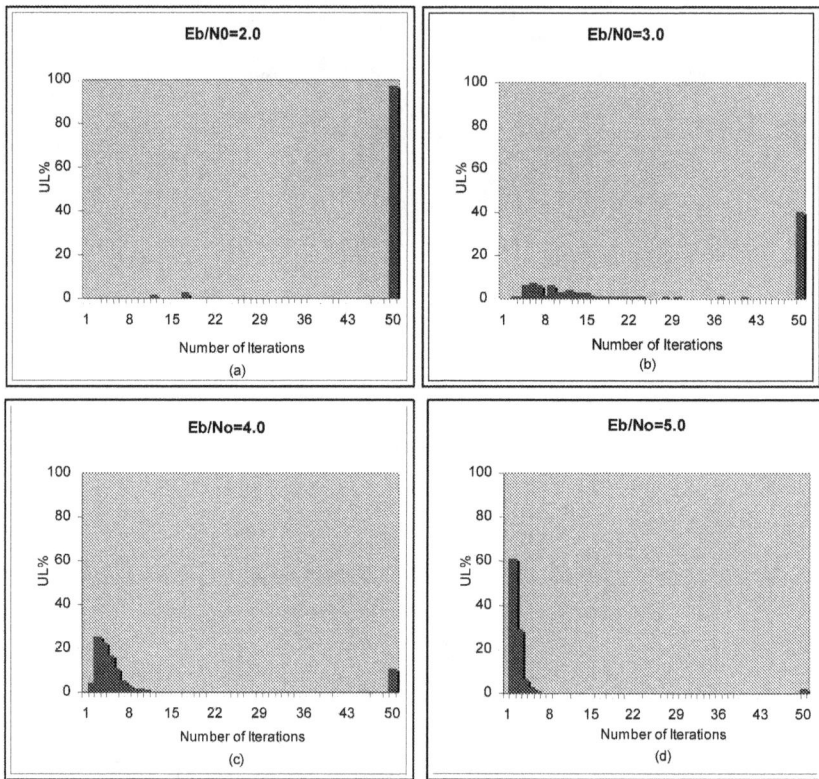

Figure 4.12: *Pdfs* of the Iterations for HS1 Algorithm with (1200, 600) Code

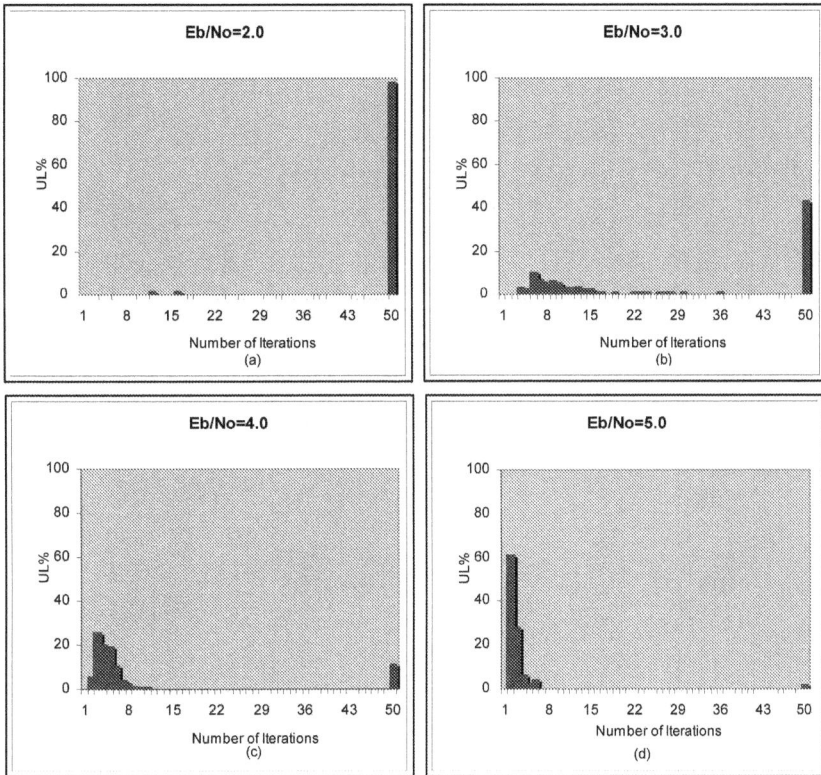

Figure 4.13: *Pdfs* of the Iterations for HS2 Algorithm with (1200, 600) Code

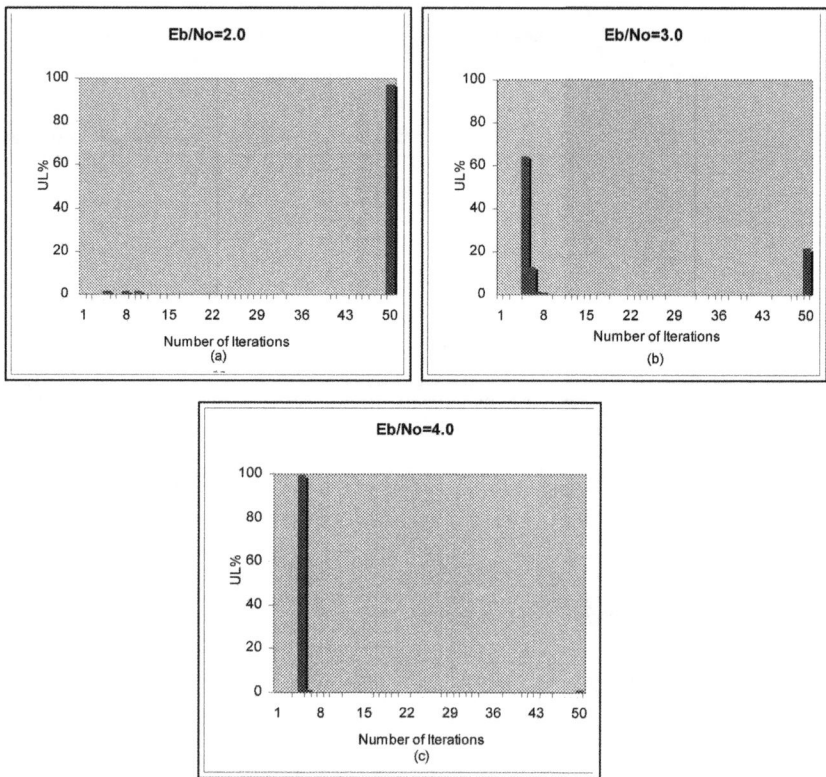

Figure 4.14: *Pdfs* of the Iterations for HS3 Algorithm with (1200, 600) Code

4.4.2 Complexity Study of (1008, 504) Irregular Code

To investigate the complexity of the decoding algorithms with irregular codes, the same trends are applied on (1008, 504) irregular LDPC code. Table 4.3 provides the statistics on required average number of iterations for convergence of decoding algorithms in their waterfall region. The statistics depend on different switching thresholds and schedules applied on algorithms.

Table 4.3: Study of the Complexity of Algorithms for (1008, 504) Code

Algorithms	Schedule	Involved Nodes	Eb/No(dB)	Average No. of Soft Iterations	Average No. of Hard Iterations	Complexity
SP	Flooding	------	1	29.5	0	Very High
			2	8.82	0	
			3	5.76	0	
GA	Flooding	------	5	0	23.2	Very Low
			6	0	3.6	
			7	0	1.8	
HS1	Deterministic	Variable/------	3	2	45.8	Low
			4	2	18.8	
			5	2	1.2	
HS2	Deterministic	Variable/Variable	3	3	25.9	Medium
			4	3	1.7	
			5	3	0.09	
HS3	Deterministic	Variable/Check	2	5	35.3	High
			3	5	2.1	
			4	5	0.03	

As an example, HS3 algorithm shows less required number of hard iterations compared to HS1 and HS2 algorithms at $Eb/No = 4$ dB. It can be explained in the way that HS3 spends more time (5 iterations) in SP algorithm compared to HS1 and HS2. Hence, more reliable information transfers into second stage and it requires less number of hard iterations to reach target bit error rate and desirable performance. In other words, it provides better performance at the cost of complexity. As another

Table 4.4: Study of the Complexity at Bit Level for (1008, 504) Code

Algorithms	Eb/No(dB)	BC_{AVE}	BC_{MAX}	Number of Edges
SP	1	236	400	4033
	2	70.5		
	3	46		
GA	5	185.6	400	4033
	6	28.8		
	7	14.4		
HS1	3	382.4	400	4033
	4	166.4		
	5	25.6		
HS2	3	231.2	400	4033
	4	38.3		
	5	24.7		
HS3	2	322.4	400	4033
	3	57.4		
	4	40.2		

example, it can be observed that the average number of hard iterations for HS1 algorithm is higher than HS2 and HS3 due to the fact that HS1 has lowest threshold value $(T = 2)$ compared to others. Hence, it spends more number of iterations in GA algorithm.

Table 4.4 represents the per bit complexity of the algorithms based on average number of iterations and maximum number of iterations for (1008, 504) code. Since the parity-check matrix of the code is not generated randomly, therefore the number of edges is constant and "E=4033". Moreover, I_{MAX} is set to 50 for all the algorithms and the dimension of the code is $K = 504$. Consequently, BC_{MAX} has the same value for all the algorithms.

Figs. 4.15-4.17 give the statistical analysis of each iteration number for irregular code in waterfall region of the algorithms. As an illustrative example, we consider Fig. 4.16 for HS2 algorithm. More than 50% of the iterations for HS2 algorithm at

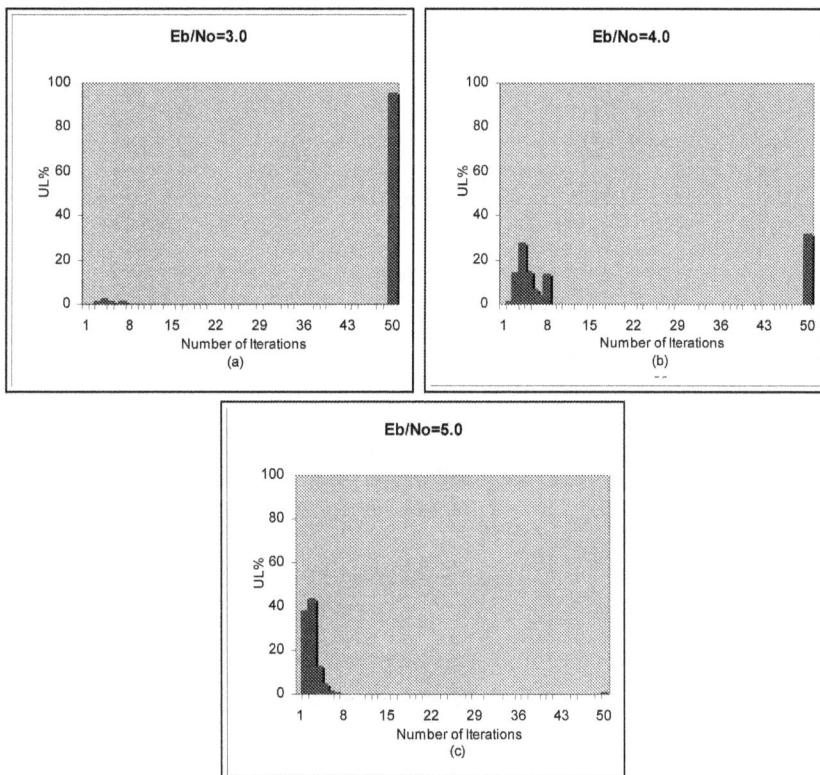

Figure 4.15: *Pdfs* of the Iterations for HS1 Algorithm with (1008, 504) Code

$Eb/No = 3.0$ dB is 50 which is the maximum number of iterations. By increasing Eb/No, the number of smaller iterations increases and their contribution becomes significant. It can be observed that at $Eb/No = 5$ dB the algorithm converges with less than 5 iterations. This trend is repeated for HS3 algorithm. Although, the speed of convergence is much faster than HS1 and HS2 in this algorithm but the average number of soft iterations and the complexity of the decoder is increased.

75

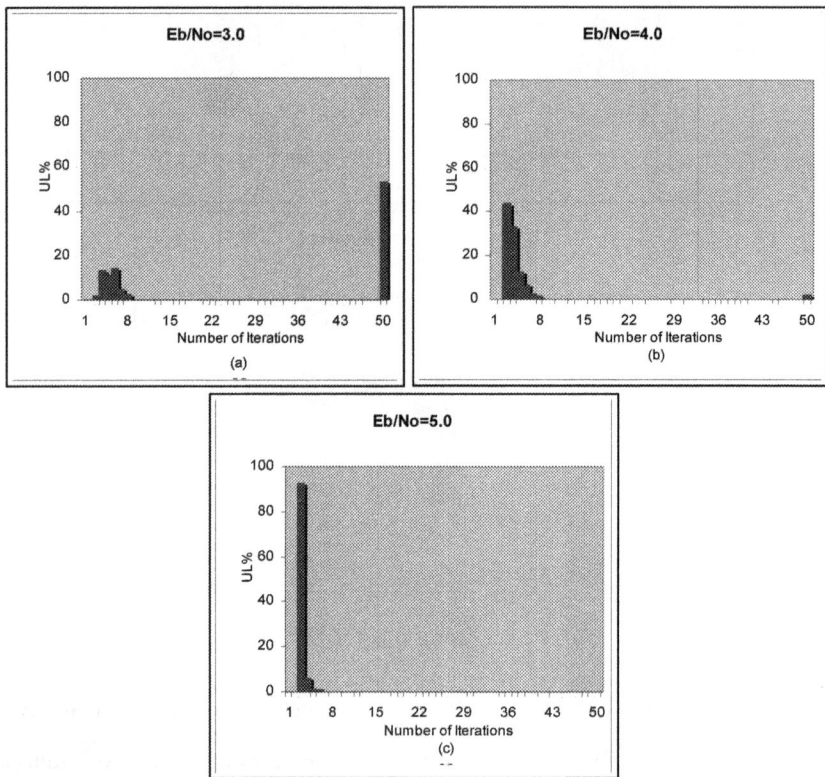

Figure 4.16: *Pdfs* of the Iterations for HS2 Algorithm with (1008, 504) Code

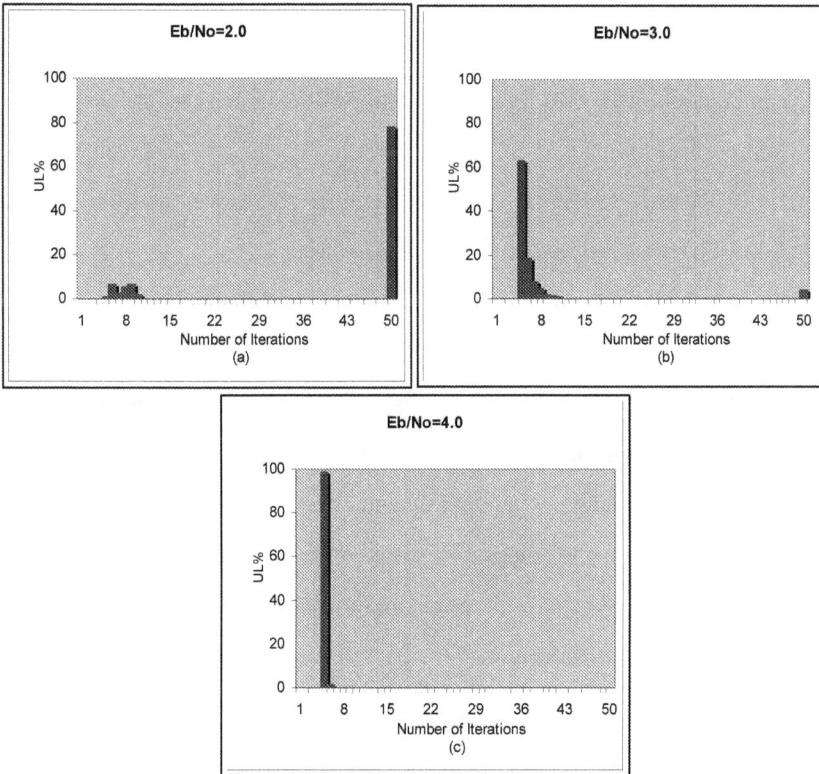

Figure 4.17: *Pdfs* of the Iterations for HS3 Algorithm with (1008, 504) Code

4.5 Summary

In this chapter, the simulation system model of the work was described and simulation results for studying the performance of the decoder were provided. The simulation results for both regular and irregular LDPC codes showed a significant improvement in performance compared to the hard-decision algorithm GA. It was shown that combination of Hybrid technique with Deterministic schedule provides a significant improvement compared to GA algorithm. It was also observed that changing the threshold value in Hybrid and selecting different involved nodes in scheduling results in various performance/complexity trade-offs.

In this chapter, the complexity of the algorithms based on number of edges and number of iterations was studied. Based on the observations, irregular codes can result in better performance with less number of iterations which results in less complexity. It was also shown that, complexity per bit for different algorithms has a direct relation with number of edges or number of connections in parity-check matrix of the code.

In conclusion, the complexity of the decoder is reduced in two ways. First, GA algorithm can not provide desirable performance with the same number of iterations. Therefore, desired performance can be obtained at higher Eb/No. On the other hand, SP provides desirable performance with the cost of complexity which is not suitable for hardware implementation. In this technique, we benefit the high performance of SP and low complexity of GA by changing the decoding rule after a few iterations. Hence, a desirable performance can be obtained with less complexity. Furthermore, since all the nodes do not update messages in each iteration, the average total number of computations is smaller than the number required for the Flooding schedule.

CHAPTER 5

CONCLUSION AND FUTURE WORK

In the first chapter of this thesis, we indicated the objectives of our work. In this chapter, the main contributions of the thesis based on our objectives are reviewed and some possible directions for future research work are suggested.

5.1 Summary of Contributions

The first objective of the work was "studying the effect of short cycles in the graph of the LDPC codes to find out *How and When* the decoding algorithm becomes suboptimal". As part of the first contribution of this work, the graph structure of LDPC codes was studied and the sub-optimality cases for decoding algorithms were highlighted. Furthermore, a graph-based search algorithm to find the shortest closed walk and shortest cycle for each node of the graph was proposed. We proved that our search algorithm is working for any desired node in the graph.

The second objective of this work was "improving the performance of the Sum-Product and Gallager A algorithms individually by preserving their optimality". As part of the second contribution of this work, we applied Deterministic schedule on variable nodes, check nodes or both variable/check nodes of LDPC regular and irregular codes in Sum-Product and Gallager A algorithms. This schedule applies the search algorithm on the nodes to find the iteration number in which the node starts to become sub-optimal. After that iteration, the node does not update messages and the optimality of the algorithm will be preserved. Moreover, Hybrid Switch-Type algorithm with different decoding thresholds was applied on improved decoding algorithms to provide a desirable performance/complexity trade-off.

The third objective of the work was "decreasing the complexity of the algorithms by reducing the average number of soft iterations and total number of computations required for convergence of algorithms". GA algorithm can not provide desirable performance at low Eb/No. On the other hand, SP algorithm provides desirable performance with the cost of complexity which is not suitable for hardware implementation. Using Hybrid technique , we benefit the high performance of soft-decision algorithms and low complexity of hard-decision algorithms by changing the decoding rule after a few iterations. Hence, a desirable performance can be obtained with less average number of soft iterations. Moreover, by applying the schedule, nodes will stop participating in update rule after losing their optimality. Therefore, all the nodes do not update messages in each iteration and the total number of computations is reduced. Consequently the requirements for the third objective are fulfilled.

The last objective of the thesis was "Designing a flexible decoder for any desired LDPC code and any combination of decoding algorithms, to provide a desirable performance/complexity trade-off based on communication systems need". The proposed technique was applied on both regular and irregular LDPC codes. Simulation results for regular random constructed (1200, 600) LDPC code with rate 1/2, degree of the columns 3 and degree of the rows 6 were given. Also, the performance of the decoder for irregular (1008, 504) optimized LDPC codes for AWGN channels were provided. In addition, a combination of soft-decision and hard-decision algorithms with different decoding thresholds were used in the decoder. The performance and complexity studies as part of this contribution proved that the designed decoder has the flexibility to work with any available LDPC code and different combination of decoding algorithms.

5.2 Future Work

The future research work can be developed in the following directions:

- Applying Hybrid Time-Invariant technique instead of Hybrid Switch-Type technique and investigating the behavior of the decoder.

- Applying the idea on large block length LDPC codes and studying the convergence behavior of the decoder.

- Combining "girth" parameter of the Deterministic schedule with other measures to design a new decoding schedule.

- Investigating the sub-optimality cases of decoding algorithms in general.

- Modifying the graph of the codes with known good parameters such as large girth and studying the behavior of the decoder on modified Tanner Graphs.

REFERENCES

[1] R. G. Gallager, "Low-Densidty Parity-Check Codes," *IEEE Transactions on Information Theory*, vol. IT-8, pp. 21–28, Jan. 1962.

[2] ——, "Low-Density Parity-Check Codes," Ph.D. dissertation, MIT Press, Cambridge MA, 1963.

[3] R. M. Tanner, "A recursive approach to low-complexity codes," *IEEE Transactions on Information Theory*, vol. IT-27, no. 5, pp. 533–547, 1981.

[4] G. A. Margulis, "Explicit constructions of graphs without short cycles and low density codes," *Combinatorica*, vol. 2, no. 1, pp. 71–78, 1982.

[5] C. Berrou, A. Glavieux, and P. Thitimajshima, "Near Shannon limit error-correcting coding and decoding: Turbo Codes," in *ICC'93 Geneva*, vol. 2, May 1993, pp. 1064–1070.

[6] D. J. C. Mackay and R. M. Neal, "Near shannon limit performance of low-density parity-check codes," *IEEE Electronic Letters*, vol. 32, pp. 1645–1646, 1996.

[7] M. Sipser and D. A. Spielman, "Expander codes," *IEEE Transactions on Information Theory*, vol. 42, pp. 1710–1722, Nov. 1996.

[8] D. J. C. Mackay, "Good error correcting codes based on very sparse matrices," *IEEE Transactions on Information Theory*, vol. 45, pp. 399–431, 1999.

[9] T. Richardson and R. Urbanke, "The capacity of Low-Density Parity-Check codes under message-passing decoding," *IEEE Transactions on Information Theory*, vol. 47, no. 2, pp. 559–618, Feb. 2001.

[10] T. Richardson, A. Shokrollahi, and R. Urbanke, "Design of capacity-approaching irregular low-density parity-check codes," *IEEE Transactions on Information Theory*, vol. 47, no. 6, pp. 619–637, Feb. 2001.

[11] N. Wiberg, "Codes and decoding on general graphs," Ph.D. dissertation, Univ. Linkoping, Linkoping, Sweden, 1996.

[12] N. Wiberg, A. H. Loeliger, and R. Kotter, "Codes and iterative decoding on general graphs," in *IEEE International Symposium on Inoformation Theory*, Whistler, Canada, Sept. 1995, p. 468.

[13] F. R. Kschischang, B. J. Frey, and H. Loeliger, "Factor graphs and the sum-product algorithm," *IEEE Transactions on Information Theory*, vol. 47, no. 2, pp. 498–518, Feb. 2001.

[14] F. R. Kschischang, "Codes defined on graphs," *IEEE Communication Mag.*, vol. 41, no. 8, pp. 118–125, Aug. 2003.

[15] K. S. Zigangirov and M. Lentmaier, "On the asymptotic iterative decoding performance of low-density parity-check codes," in *Proc. Int. Symp. on Turbo codes and related topics*, Brest, France, Sept. 2000, pp. 39–42.

[16] G. D. Forney, "On iterative decoding and the two-way algorithm," in *Int. Symp. on Turbo codes and related topics*, Brest, France, Sept. 1997, pp. 12–25.

[17] A. Anastasopoulos, "A comparison between the sum-product and the min-sum iterative detection algorithms based on density evolution," in *Proc. IEEE Global Telecommunication Conference (Globecom)*, vol. 2, Nov. 2001, pp. 1021–1025.

[18] J. Zhao, F. Zarkeshvari, and A. H. Banihashemi, "On implementation of min-sum algorithm and its modifications for decoding low-density parity-check (LDPC) codes," *IEEE Transactions on Communications*, vol. 53, no. 4, pp. 549 – 554, 2005.

[19] Y. Kou, S. Lin, and M. P. C. Fossorier, "Low-density parity-check codes construction based on Finite Geometries : a rediscovery and new results," *IEEE Transactions on Information Theory*, vol. 47, pp. 2711–2736, Nov. 2001.

[20] J. Zhang and M. Fossorier, "Shuffled belief propagation decoding," in *Signals, Systems and Computers, 2002*, Nov. 2002.

[21] M. M. Mansour and N. R. Shanbhag, "Turbo decoder architectures for low-density parity-check codes," in *IEEE Global Telecommunication Conference (GLOBECOM)*, Nov. 2002.

[22] Y. Mao and A. H. Banihashemi, "Decoding low-density parity-check codes with probabilistic schedule," *IEEE Communications Letters*, vol. 5, no. 10, pp. 414–416, Oct. 2001.

[23] ——, "A new schedule for decoding low-density parity-check codes," in *IEEE Globecom 2001, Texas*, vol. 2, Nov. 2001, pp. 1007–1010.

[24] ——, "Decoding low-density parity-check codes with probabilistic schedule," in *Proc. IEEE PACRIM 2001, Victoria, Canada*, vol. 1, Aug. 2001, pp. 119–123.

[25] A. Nouh and A. H. Banihashemi, "A new decoding algorithm for low-density parity-check codes," in *21st. Biennial Symposium on Communications, Queen's University, Kingston, Canada*, June 2002.

[26] ——, "Reliability-based schedule for bit-flipping decoding of low-density parity-check codes," *IEEE Transactions on Communications*, vol. 52, no. 12, pp. 2038–2040, Dec. 2004.

[27] H. Xiao and A. H. Banihashemi, "New schedules for decoding LDPC codes," in *21st. Biennial Symposium on Communications, Queen's University, Kingston, Canada*, June 2002.

[28] ——, "Graph-based message-passing schedules for decoding LDPC codes," *IEEE Transactions on Communications*, vol. 52, no. 12, pp. 2098 – 2105, Dec. 2004.

[29] P. Zarrinkhat and A. H. Banihashemi, "Hybrid decoding of low-density parity-check codes," in *Proc. Int. Symp. on Turbo codes and Related Topics*, Brest, France, Sept. 2003, pp. 503–506.

[30] ——, "Hybrid hard-decision iterative decoding of regular low-density parity-check codes," in *IEEE ICCC 2004*, Paris, France, 2004, pp. 435–439.

[31] ——, "Hybrid hard-decision iterative decoding of regular low-density parity-check codes," *IEEE Communications Letters*, vol. 8, no. 4, pp. 250–252, Apr. 2004.

[32] ——, "Hybrid decoding of irregular low-density parity-check codes," in *IEEE ISIT 2005*, Adelaide, Australia, Sept. 2005, pp. 312–316.

[33] M. Ardakani and F. R. Kschischang, "Gear-Shift Decoding," in *21st. Biennial Symposium on Communications, Queen's University, Kingston, Canada*, June 2002.

[34] ——, "Gear-Shift decoding for algorithms with varying complexity," in *IEEE ICC 2005*, vol. 1, Seoul, Korea, May 2005, pp. 500–504.

[35] ——, "Gear-shift decoding," *IEEE Transactions on Communications*, vol. 54, no. 7, pp. 1235–1242, 2006.

[36] A. J. Felstrom and K. S. Zigangirov, "Time-varying periodic convolutional codes with low-density parity-check matrix," *IEEE Transactions on Information Theory*, vol. 45, pp. 2181–2191, 1999.

[37] M. C. Davey and D. J. C. Mackay, "Low density parity check codes over GF(q)," *IEEE Communications Letters*, vol. 2, pp. 159–166, 1998.

[38] M. Luby, M. Mitzenmacher, M. Shokrollahi, and D. Spielman, "Improved low-density parity-check codes using irregular graphs," *IEEE Transactions on Information Theory*, vol. 47, pp. 585–598, Feb. 2001.

[39] R. Zhang, "Linear block codes," Department of Electrical and Computer Engineering, Drexel University, Tech. Rep. ECE-S622/T602, Fall 2002.

[40] J. Compello, D. S. Modha, and S. Rajagopalan, "Designing LDPC codes using bit-filling," in *ICC 2001*, June 2001, pp. 55–59.

[41] Y. Mao and A. H. Banihashemi, "A heuristic search for good LDPC codes at short block lengths," in *ICC 2001*, June 2001, pp. 41–44.

[42] Y. Kou, S. Lin, and M. P. C. Fossorier, "Low-density parity-check codes construction based on finite Geometry," in *Globecom2000*, Nov. 2000, pp. 825–829.

[43] H. Song, J. Liu, and B. V. K. V. Kumar, "Low-density parity-check codes for partial response channels," in *Globecom 2002*, Nov. 2002, pp. 1294–1299.

[44] D. J. C. Mackay, S. T. Wilson, and M. C. Davey, "Comparison of constructions of irregular Gallager codes," *IEEE Transactions on Communications*, vol. 47, pp. 1449–1454, Oct. 1999.

[45] T. J. Richardson and R. L. Urbanke, "Efficient encoding of low-density parity-check codes," *IEEE Transactions on Information Theory*, vol. 47, pp. 638–656, Feb. 2001.

[46] T. H. Cormen, C. E. Leiserson, R. L. Rivest, and C. Stein, *Introduction to Algorithms*. Cambridge: MIT Press, 1990.

[47] J. L. Gross and J. Yellen, *Graph theory and its applications*. Chapman and Hall, 2006.

[48] T. Tian, C. Jones, J. D. Villasenor, and R. D. Wesel, "Selective avoidance of cycles in irregular LDPC code construction," *IEEE Transactions on Communications*, vol. 52, no. 8, pp. 1242–1247, Aug. 2004.

[49] T. R. Halford and K. M. Chugg, "An Algorithm for Counting Short Cycles in Bipartite Graphs," *IEEE Transactions on Information Theory*, vol. 52, no. 1, pp. 287–292, Jan. 2006.

[50] D. Mackay. Encyclopedia of sparse graph codes. [Online]. Available: http://www.inference.phy.cam.ac.uk/mackay/codes/data.html

[51] J. Kim, U. Peled, I. Perepelitsa, V. Pless, and S. Frieldland, "Explicit construction of families of LDPC codes with no 4 cycle," *IEEE Transactions on Information Theory*, vol. 50, pp. 2378–2388, Oct. 2004.

[52] M. Baldi, G. Bosco, F. Chiaraluce, and R. Garello, "Decoding Complexity and Iteration Number Statistics in Low Density parity Check Codes," *Proceedings of the 4th Int. Symposium on Information and Communication Technologies*, vol. 92, pp. 81–86, 2005.

www.ingramcontent.com/pod-product-compliance
Lightning Source LLC
Chambersburg PA
CBHW082119080925
32270CB00071B/6516